Personal Data-Smart Cities: How Cities Can Utilise Their Citizen's Personal Data to Help Them Become Climate Neutral

RIVER PUBLISHERS SERIES IN ENERGY SUSTAINABILITY AND EFFICIENCY

Series Editor

PEDRAM ASEF
Lecturer (Asst. Prof.) in Automotive Engineering,
University of Hertfordshire,
UK

The "River Publishers Series in Energy Sustainability and Efficiency" is a series of comprehensive academic and professional books which focus on theory and applications in sustainable and efficient energy solutions. The books serve as a multi-disciplinary resource linking sustainable energy and society, fulfilling the rapidly growing worldwide interest in energy solutions. All fields of possible sustainable energy solutions and applications are addressed, not only from a technical point of view, but also from economic, social, political, and financial aspects. Books published in the series include research monographs, edited volumes, handbooks and textbooks. They provide professionals, researchers, educators, and advanced students in the field with an invaluable insight into the latest research and developments.

Topics covered in the series include, but are not limited to:

- Sustainable energy development and management;
- Alternate and renewable energies;
- Energy conservation;
- Energy efficiency;
- Carbon reduction;
- Environment.

For a list of other books in this series, visit www.riverpublishers.com

Personal Data-Smart Cities: How Cities Can Utilise Their Citizen's Personal Data to Help Them Become Climate Neutral

Editors

Shaun Topham

DataVaults Project, Ireland

Paolo Boscolo

Comune of Prato and Major Cities of Europe

Michael Mulquin

Open and Agile Smart Cities

River Publishers

Routledge
Taylor & Francis Group
LONDON AND NEW YORK

Published 2022 by River Publishers
River Publishers
Alsbjergvej 10, 9260 Gistrup, Denmark
www.riverpublishers.com

Distributed exclusively by Routledge
4 Park Square, Milton Park, Abingdon, Oxon OX14 4RN
605 Third Avenue, New York, NY 10017, USA

Personal Data-Smart Cities: How Cities Can Utilise Their Citizen's Personal Data to Help Them Become Climate Neutral / Shaun Topham, Paolo Boscolo and Michael Mulquin.

Routledge is an imprint of the Taylor & Francis Group, an informa business

ISBN 978-87-7022-800-8 (print)
ISBN 978-87-7022-799-5 (ebook master)

The cover design is based upon the graphic recordings produced by Agata Smok, during the EU Smart Cities MarketPlace Action Cluster meetings.
©2021 Graphic recording by Agata Smok for DG ENER
https://smart-cities-marketplace.ec.europa.eu/

Contents

Foreword

Johan Bodenkamp, Policy and Programme Officer at the European Commission, and DataVaults Project Officer

It is widely acknowledged that data has an enormous potential to contribute to better-informed decisions and to create new services and products for the benefit of all.

Every day, we generate an unprecedented amount of data, which is increasingly harvested and used to foster innovation and sustainable economic growth as well as to improve the overall wellbeing of society. Despite the current sanitary, economic, and geopolitical context, studies confirm this positive trend in the data economy. Having reached 440 billion euros in 2021, the latest European Data Market Study[1] estimates that the value of the EU27 data economy will reach 600 billion euros in 2025 and 1 trillion euros in 2030. It is, therefore, not surprising that stakeholders regularly refer to data as the new oil or electricity of our modern society.

The EU is fully committed to setting a framework to help ensure that the benefits of the ongoing data revolution are fully reaped. To this end, the European data strategy[2] aims to make the EU a leader in a data-driven society and to create a single European data space. This space would be a genuine European single market for data, open to data from across the world, in which an almost infinite amount of high-quality personal and non-personal data is secure and can be used in an ethical and trusted way to create value and boost growth, while minimising the environmental footprint related to its use.

To create the right conditions and gradually realise this bold vision for Europe, ambitious legislative, research, innovation, and deployment actions at EU and national level are underway. In particular, the recently adopted Data Governance Act[3] and the proposed Data Act[4] are key EU legislative measures to enhance data availability and use in line with EU rules and values.

Despite this positive trend, the use and sharing of personal data is still limited. This is partly due to hesitations on how to benefit from such sensitive data in a trusted and secure way, ensuring respect of data protection legislation and allowing data subjects and data holders to remain in control of their data and its use. Fortunately, several EU-funded and national projects, some of which have been actively involved in the preparation of this work, concretely show how this can be done through the use of appropriate secure and privacy-preserving techniques.

This book provides valuable insights into how smart cities could use citizens' personal data in a trusted and ethical way. It makes it an extremely useful source of information for city and municipal staff as well as for other interested stakeholders.

The hope is that it will inspire smart cities to engage more actively in using relevant data, and in particular citizens' personal data, to support important local policy objectives, notably to become climate neutral as quickly as possible.

Brussels,
July 20 2022

Preface

This book was produced in order to help grow a data economy in Europe, which reflects European values. It is argued that by adding a citizen's personal data, under their own control, to current and developing use of data in a smart city, a major contribution can be made to realising the ambition of many European cities of becoming carbon-neutral by 2030. And further, to contribute to building a mechanism for replicating the lessons which will be learned as cities utilise personal data and progress towards achieving both their environmental and smart city targets.

Acknowledgements

European Commission

An initiative of the Smart Cities Marketplace #EUSmartCities

Funded by the European Union

The 16 projects listed below have all received funding from the European Union's Horizon 2020 research and innovation programme under the respective Grant Agreements indicated.

DataVaults This project has received funding from the European Union's Horizon 2020 research and innovation programme under grant agreement No 871755

EUHubs4Data project has received funding from the European Union's H2020 Research and Innovation Programme under the Grant Agreement No. 951771.

KRAKEN is a project which has received funding from the European Union's Horizon 2020 (H2020) research and innovation programme under the Grant Agreement No. 871473.

DataPorts
Data Platform for the Connection of Cognitive Ports

DataPorts project has received funding from the European Union's Horizon 2020 Research and Innovation Programme under Grant Agreement No. 871493.

Safe-DEED

Safe-DEED is a project which has received funding from the European Union's Horizon2020 research and innovation programme under grant agreement No. 825225.

REPLICATE

Replicate project has received funding from the European Union's Horizon 2020 research and innovation programme under Grant Agreement No. 691735.

RUGGEDISED
Designing smart, resilient cities for all

RUGGEDISED is a project that has received funding from the European Union's Horizon 2020 research and innovation programme under grant agreement No. 731198.

PIM City

PIMCity project has received funding from the European Union Horizon 2020 Research and Innovation programme under the ICT theme: ICT-13-2018-2019 Grant Agreement No. 871370.

InteropEHRate
EHR in people's hands across Europe

InteropEHRate project has received funding from the European Union's Horizon 2020 research and innovation programme under grant agreement No. 826106.

smashHit

SmashHit project has received funding from the European Union's Horizon 2020 research and innovation programme under Grant Agreement No. 871477.

DUET

DUET is a project has received financial support from the European Union's Horizon 2020 Programme under grant agreement No. 870697.

PolicyCloud has received funding from the European Union's Horizon 2020 research and innovation programme under grant agreement No. 870675.

i3-Market project is funded by the European Union's H2020 Research and Innovation Programme under the Grant Agreement No. 871754 (i3-MARKET).

IRIS project has received funding from the European Union's Horizon 2020 research and innovation programme under grant agreement No. 774199.

Auroral project has received funding from the European Union's Horizon 2020 research and Innovation Programme under grant agreement No. 101016854.

SmartenCity project has received funding from the European Union's Horizon 2020 research and innovation programme under grant agreement No. 691883

Major Cities of Europe[5]

Major Cities of Europe is a European independent organisation of local government CIOs, heads of departments, and policy-makers. They co-operate with academia, public and private organisations focused in that domain. They promote the development of new services for citizens and local businesses. They facilitate exchanges of strategies, experiences, ideas, and

[5] https://www.majorcities.eu/

solutions for the use of information and communication technology. Many of their members are also amongst those 100 cities selected by the Commission to lead, in becoming climate-neutral by 2030.

Open & Agile Smart Cities[6]

Open and Agile Smart Cities bring together smart cities and communities worldwide to shape the global market for digital services. It is a network that connects cities and communities worldwide to learn from each other and exchange digital, data-driven solutions based on minimal interoperability mechanisms (MIMs). They are creating sustainable impact for their member cities by working towards a common technical ground for cities and communities – based on open standards, open APIs, and shared data models. As with Major Cities, many members are seeking to be climate-neutral by 2030.

Disclaimer

The information and views set out in this publication are those of the author(s) and do not necessarily reflect the official opinion of the European Communities. Neither the European Union institutions and bodies nor any person acting on their behalf may be held responsible for the use which may be made of the information contained therein.

We can confirm that there are no conflicts of interest to report.

---

[6] https://oascities.org/

List of Contributors

Alisa Jashari, *Tampere, Finland*

Ana García Garre, *MIWenergia, Spain*

Annina Autero, *SURE Project, Tampere, Finland*

Borja Molina-Ríos, *MIWenergia, Spain*

Christina N. Karaberi, *e-trikala – Department of Research and Communication, Trikala, Greece*

Christina Tsiligkiri, *Olympiacos SFP, Pireas, Greece,*

Elena Palmisano, *Comune di Prato, Italy*

Elena Palmisano, *PIN, Prato, Italy*

Emanuele Geri, *Information Systems Directorate, City of Florence, Italy*

Esben Pejstrup, *ICLEI, Germany*

Fabio Pazzaglia, *Information Systems Directorate, City of Florence, Italy*

Haydee S. Sheombar, *Founder Kankan*Tree, Research Affiliate, Erasmus University Rotterdam, Netherlands*

Hervé Groléas, *Chief Digital and Information Officer, Grand Lyon, France*

Ilesh Datani, *Assentian, Ireland*

Iván Carrillo Quero, *ATOS, Spain,*

John Kaldis, *Unisystems, Greece*

John Ketchell, *European Association for Coordinating Consumer Representation in Standardisation (ANEC)*

Joseph V. Sheombar, *Master Student Built Environment, Delft University of Technology, Netherlands*

Juergen Ritzek, *EEIP, Brussels, Belgium*

Maria Sangiuliano, *Research Director, Smart Venice, Venice, Italy*

Marina Da Bormida in Cugurra, *Genoa, Italy*

Marina Micheli, *Scientific Project Officer, Digital Economy Unit, European Commission's Joint Research Centre (JRC), Ispra, Italy*

María Jose Lopez Osa, *Tecnalia, Spain*

Michael Mulquin, *MIMs Ambassador, Open & Agile Smart Cities*

Michail Bourmpos, *IT Department, City of Piraeus, Greece*

Mihnea Tufiş, *Eurecat Technology Center of Catalunya, Carrer de Bilbao, Barcelona*

Minna Sapyska, *Tampere, Finland*

Paolo Boscolo, *Comune di Prato, Italy*

Sari Maenpaa, *Insta, Finland*

Shaun Topham, *Lead, EU Smart Cities Marketplace, "Citizen's Control of Data" Initiative*

Sotiris Koussouris, *Suite5, Cyprus*

Vincent Keunen, *Andaman7, Belgium*

Yury Glikman, *Fraunhofer, Germany*

List of Figures

List of Tables

List of Abbreviations

ABAC	Attribute-based access control
ABE	Access-based enumeration
AI	Artificial intelligence
API	Application programming interface
B2G	Business to government
CERTs	Computer emergency response (or readiness) team
CIM	City innovation models
CRUD	Create, read, update, and delete
CSIRT	Computer security incident response team
DGA	Data governance act
DLT	Distributed ledger technology
DMA.	Digital markets act
DQA	Data quality assessment
DQD	Data quality dimension
DQM	Data quality metric
DSA	Digital services act
DVC	Data valuation component
DVP	Data valuation process
EEA	European economic area
eIDAS	Electronic identification, authentication and trust services
EOSC	European open science cloud
ESCO	Energy service company
EU	European union
FAIR	Findable-accessible-interoperable-reusable
GDPR	The general data protection regulation
GIS	Geographic information system
GPS	Global positioning system
HMRC	Her majesty's revenue and customs service
ICT	Information and communications technology
IoT	Internet of things

ITU	International telecommunication union
JRC	Joint research centre
MIM	Minimal interoperability mechanisms
NSO	National statistics organisation
PDV	Personal data vault
PIMS	Privacy information management system
PSI	Public sector information
PV	Photovoltaics
R&D	Research and development
SME	Small- and medium-sized enterprise
SSI	Self-sovereign identity
TPM	Trusted platform module
UDP	Urban data platform
USD	United states dollar
VAT	Value added tax
WIMAX	Worldwide interoperability for microwave access

Introduction

Shaun Topham[1], Paolo Boscolo[2], Michael Mulquin[3]

[1]Initiative Lead, "Citizen Control of Personal Data", EU Smart Cities MarketPlace
[2]City of Prato, Initiative Co-Lead and "Major Cities of Europe"
[3]MIMs Ambassador, "Open & Agile Smart Cities"

The advantages of utilising a citizen's personal data within a smart city context are the primary focus for this work. At the same time, we recognise that the general usage of all data in a smart city, in a joined-up way, is faced with obstacles hindering its full exploitation, and, naturally, these obstacles will hamper the take-up of the emerging technologies that are looking to release a citizen's personal data into this ecosystem. However, the current and extensive range of projects featured in this work will go a long way to overcoming those already identified obstacles, which are currently hindering the spread of "urban data platforms" across Europe.

One of the main results that we would envisage from putting this book together is to assemble a key group of opinion-forming cities, mainly from those recently accepted as the 100 cities in the EU Cities Mission.[1,2] This group should act as a "lighthouse" to first determine the approach to adopting the emerging technologies from EU-funded research and, second, how

[1] "EU Mission: Climate-Neutral and Smart Cities - European Commission." https://ec.europa.eu/info/research-and-innovation/funding/funding-opportunities/funding-programmes-and-open-calls/horizon-europe/eu-missions-horizon-europe/climate-neutral-and-smart-cities_en (accessed Jul. 25, 2022).

[2] European Commission, Directorate-General for Research and Innovation, Gronkiewicz-Waltz, H., Larsson, A., Boni, A., et al., "100 climate-neutral cities by 2030 - by and for the citizens : report of the mission board for climate-neutral and smart cities", Publications Office, 2020, https://data.europa.eu/doi/10.2777/46063

DOI: 10.1201/9781003399384-1

personal data might be included to enhance the whole smart city approach to data handling. And as a consequence, this will lead to having a clear sense of how utilising personal data may contribute to the goal of becoming smart and climate-neutral by 2030, and, subsequently, help to steer the rest of the cities towards achieving this goal by 2050.

The advantages of utilising citizen's personal data in the cities will be demonstrated, both as exemplars from the work carried out in EU-funded research projects and also in the way it can complement the existing good use of data in a smart city.

The European data economy requires a pathway to adoption at scale. Around 10% of the EU population live in the first 100 cities and if you add those who live in the further 250 cities having a similar goal and then add to that other follower cities, a critical mass is well within sight.[3,4]

Chapters covering the legal landscape and covering governance issues pave the way to creating an ecosystem involving cities and their stakeholders to capitalise on the evolution of the European Data model, as the re-addressing of the current balance of power shifts value in the usage of personal data back towards Europe and away from the current global situation. This model is steering away from the existing situation which sees much data flowing one way, along with associated revenue generation, to outside of the EU. "Some large online platforms act as "gatekeepers" in digital markets. The Digital Markets Act aims to ensure that these platforms behave in a fair way online. Together with the Digital Services Act, the Digital Markets Act is one of the centrepieces of the European digital strategy".[5]

What needs to be considered in the grand scheme of things is "If the revenues start to remain more in Europe, how can we make sure that some of this value generated remains in local smart city ecosystems, contributing to achieving the goals of achieving climate-neutrality and for social good, rather than to the balance sheets of some of the world's largest corporations?"

We will scrutinise how best to put value on the data which is available and, as a non-rival good, can be used over and over again for multiple purposes.

---

[3] "Building a data economy — Brochure: Shaping Europe's digital future." https://digital-strategy.ec.europa.eu/en/library/building-data-economy-brochure (accessed Jul. 25, 2022).

[4] "A European Strategy for data. Shaping Europe's digital future." https://digital-strategy.ec.europa.eu/en/policies/strategy-data (accessed Jul. 25, 2022).

[5] "The Digital Markets Act: ensuring fair and open digital markets - European Commission." https://ec.europa.eu/info/strategy/priorities-2019-2024/europe-fit-digital-age/digital-markets-act-ensuring-fair-and-open-digital-markets_en (accessed Jul. 25, 2022).

This book sets out to address some of the issues which a smart city needs to overcome in order to make use of the data currently available to them and draws attention to how a citizen's personal data can be accessed technically in order to take existing best practice to the next level, and, in particular, as to how the technology emerging which safely allows a citizen to share their personal data, under their own control, can be utilised to achieve these goals.

It is aimed at providing answers for those working within a smart city, who need to advice their mayors or leaders with regard to introducing new technology. The target group is primarily those leaders and strategists from a smart city who take the decisions, and, in particular, the new grouping of 100 "net zero carbon" cities,[6] of which over 35 are linked to either "Major Cities of Europe" or "Open & Agile Smart Cities" with many others linked to the MarketPlace.[7]

But, they do not take the decision in isolation, and so specific topics of concern to the legal teams, service providers, urban planners, technical teams, and, importantly, the financial people will be addressed, so that there are no remaining impediments to taking a positive decision to utilise these evolving technologies. Whoever took the actual decision to become a climate-neutral city will need support. Experience with previous initiatives aimed at sharing best practice has shown that this is more likely to be achieved when peers talk to peers.

When "Big Data" first appeared on the scene, there was a strong interest developed regarding how it may be of benefit to a smart city. However, an issue was that there was a lack of good examples of big data in use, shown in an easily recognisable way for those running a city. They were difficult to find. We hope to provide a range of examples of how accessing a citizen's personal data, with all the safeguards in place, can enhance the work a city is already implementing and using the data it already has access to. These exemplars will come from a raft of projects and from existing best practice.

Again, brief descriptions of these projects will be provided in order to understand the context of their contributions.

There will be simple and more readily understood coverage of the technology which is being produced but with more in-depth supporting papers

---

[6] European Commission, Directorate-General for Research and Innovation, EU missions: 100 climate-neutral and smart cities, Publications Office of the European Union, 2022, https://data.europa.eu/doi/10.2777/191876

[7] https://smart-cities-marketplace.ec.europa.eu/

being sign-posted or being made available in a complementary resource to be hosted within the EU's Smart Cities Marketplace domain.[8]

Aspects of the use of citizen's personal data can also make their own direct contribution to driving forward the "carbon-neutral" agenda, particularly with regard to increasing citizen participation in the decision-making process.

Case studies of existing best practice in the use of data by smart cities are augmented with examples of how, by also embracing a citizen's personal data in the mix, this will enable better services to be created and potential new revenue streams to occur. This aspect will be significant because, if handled correctly, and new eco-systems emerge, with suitable governance and legal and ethical basis, this will enable new business models and investment opportunities to emerge, combining the original data applications with new personal data.

What is clear is that the more applications available to share between cities, the greater likelihood there is of success, and that by striving for interoperability between all the initiatives taking this agenda forward, the odds are further reduced.

In pursuit of this objective, we will attempt to "tell the story of data" within a city to help focus interest on what needs to be done next. It will provide a building block in understanding the value of this data in a local context, thus starting to remove another significant obstacle of how to decide whether to invest or not.

We will conclude by looking at what new technologies will be emerging in the coming years from the projects about to start. By shortening the gap from research to deployment, the hundred cities with carbon-neutral targets by 2030 will have more chance of succeeding.

Thus, "Personal Data-Smart Cities" sets out to establish how the results of both existing EU-funded projects and other future projects already in the pipeline can help cities utilise their citizen's personal data in achieving the goal of becoming climate neutral by 2030, but also to enable closer engagement with their citizens, whilst contributing to the development of a European model for the data economy.

[8] "Citizen's Control of Personal Data - Smart Cities Marketplace." https://smart-cities-marketplace.ec.europa.eu/action-clusters-and-initiatives/action-clusters/citizen-focus/citizens-control-personal-data (accessed Jul. 25, 2022).

1

Peril on the Road to Utopia – Opportunities and Risks of Infusing Personal Data into the Smart City Ecosystem

Haydee S. Sheombar[1] and Joseph V. Sheombar[2]

[1]Founder Kankan*Tree, Research Affiliate, Erasmus University Rotterdam, Netherlands
[2]Master Student Built Environment, Delft University of Technology, Netherlands
Contact: sheombar@rsm.nl

Abstract

Technology is not always living up to its promises. The idea of smart cities has been around for over ten years and is finally gaining traction. Urban data platforms are being adopted across Europe. However, smart cities still struggle to engage citizens. Also struggling is the internet. Many, including the founder of the World Wide Web, feel that it is broken: hijacked by big tech, trapping us into filter bubbles. How can the use of personal data remedy both of these struggles? With control over our personal data using personal data vaults, we could limit exposure to unwanted manipulation online. Next, by rendering our personal data for solving societal problems, we help improve smart city services. Finally, we can use insights from our integrated personal data to change our own behaviour and become truly smart citizens. Sounds too good to be true? Maybe. But more importantly, let us not repeat past mistakes and, this time, put in place measures to identify and resolve adverse effects. EU legislation and European values are foundational measures. In addition, by using scenario thinking, we propose two broad strategies for finding more measures. One is adopting a human-centric paradigm, rather than an economic paradigm. The other is adopting a contextual-values paradigm rather than an individualistic paradigm. We recommend that the use of

DOI: 10.1201/9781003399384-2

personal data should not be left to technocrats with an exploitative mind set. It should be handled by city designers that can indeed imagine utopia and craft a path towards it.

1.1 Introduction

In Greek mythology, Daedalus, a skilled inventor, wanted to help his son Icarus escape captivity. He built him wings made of feathers held together with wax. In his enthusiasm, Icarus does not heed his father's warning and flies too close to the sun. The wax melts, and Icarus falls to his death in the Aegean Sea.

This myth teaches us that technology can "give us wings" but that we need to handle it with care. We should be well aware of our own limitations and those of technology. But, most importantly, we should not fall prey to hubris, like Icarus did.

Our world is digitalising at a vast pace and, today, almost every aspect of our lives can be captured in data. Can these *personal data* be used to make the world a better place? And how do we know when we are "flying too close to the sun"? These are the questions that inspired this essay.

1.2 Broken Promises

Can you imagine a world without digital technologies? They are pervasive, make our lives and our world hyper-connected and hyper-intelligent, and blur the line between the physical and virtual. You do not have to be a Luddite to have real concerns about the expectations and the use of digital technologies by companies, governments, and citizens. Let us have a critical look at the idea of a smart city that promises to make our cities liveable, and at the Internet that would enable us all, promising a more equal and democratic society.

1.2.1 The smart city is finally coming of age

With the vast urbanisation at a global scale, many of today's ecological and socio-economic challenges concentrate in urban areas. The technology companies like IBM and Cisco that minted the term smart cities more than a decade ago, therefore, thought that cities were a good place to start solving some of our societal problems by using digital technology.

The first cities emerged only 6000 years ago in Mesopotamia. Cities rise and fall and their demise is caused by natural disaster, changing economics,

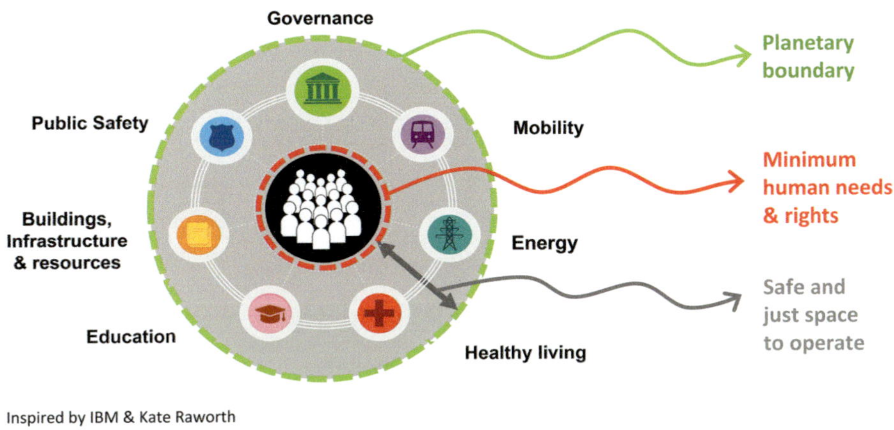

Figure 1.1 A smart city as a bounded "system of systems".

war, or environmental depletion. The latter was the case for Tikal, one of the great cities of Mayan civilisation. Back then and still today, not crossing planetary or ecological boundaries is essential. With (digital) technology, we hope to prevent looming demise on a much grander scale than Tikal. A smart city should help us stay within planetary boundaries, whilst catering to minimum human needs and fundamental human rights (see Figure 1.1).

There are at least two signs that the concept of smart cities is finally gaining traction. The first one is the fact that of the 80 cities surveyed over two years ago as part of the Horizon 2020 RUGGEDISED project,[1] 30% had an operational Urban Data Platform, and all of the rest were either in the exploration or planning phase. Urban Data Platforms are increasingly seen as the vital infrastructure that can help smart city pilots scale. Another sign is that the EU believes that the way to fight climate change is through a "Twin Transition", i.e., the energy transition and the digital transition, together. This twin transition is also embedded in "The EU Smart Cities Mission" that aims to have 100 smart cities that will lead the way to climate neutrality.

Two findings from the RUGGEDISED study stand out. First, of all, trust was identified as one of the critical success factors for an Urban Data Platform. Trust in the technology and capabilities, trust in the organisations that control

[1] Sheombar, H., Van Oosterhout, M., Diran, D., Bagheri, S., & Popp Larsen, C. (2020, November). Governance, Trust and Smart City Business Models: the Path to Maturity for Urban Data Platforms (RUGGEDISED – 731198). EUROPEAN COMMISSION. https://ruggedised.eu/fileadmin/repository/Publications/RUGGEDISED-D6.6-Governance-Trust-SmartCity_business_Models-EUR-FINAL-2020.11.13.pdf

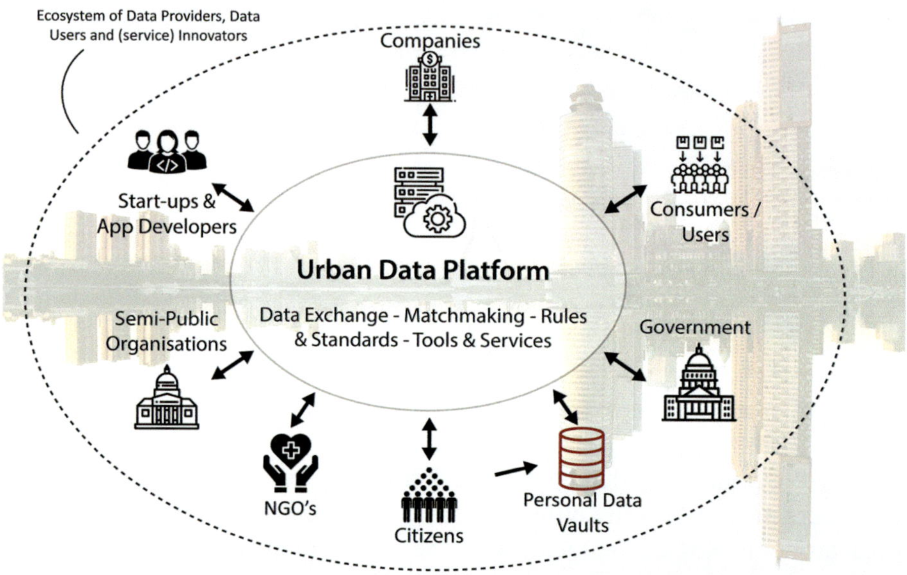

Figure 1.2 The Urban Data Platform, personal data vaults and the smart city ecosystem (based on RUGGEDISED).

the platform, and, most importantly, mutual trust between the public and private sector involved in financing, owning, and managing the platform. The study also showed that cities find it really hard to engage citizens: everyone says that it is important; hardly anyone is actually doing it. This is not really surprising because the smart cities approach to date was skewed towards smart infrastructure, whilst a smart citizen is an integral part of the smart city.

Not only are smart cities about the twin transition, they are also characterised by "twin speed". As Townsend recalls in his book about smart cities, it took a year to build the congestion charging system for the city of London in 2002, while it took policy- and law-makers almost 40 years of deliberation to decide.[2] This difference in speed between technology and policy may explain the slow progression of the smart cities. Another explanation is the low involvement of citizens. Using citizen's personal data by means of personal data vaults (PDVs) (see Figure 1.2) might be a way to improve citizen engagement in the smart city ecosystem.

[2] Townsend, A. M. (2013). Smart cities: Big data, civic hackers, and the quest for a new utopia. WW Norton & Company.

1.2.2 Is the internet broken?

In March 1989, Tim Berners-Lee, a CERN employee at the time, wrote a proposal for what would become the World Wide Web. He proposed a web of "hypertext documents" that could be viewed by "browsers", making it easier for CERN researchers to find and share their research outcomes.[3] Soon after its inception, the World Wide Web became the most important service on the internet.[4] Berners-Lee's original vision for the web was a medium for secure and decentralised exchange of data.[5] He hoped it would become an open and free environment with egalitarianism at its core.

Fast forward 30 years, and this dream has been shattered. In 2019, Berners-Lee called out that "Platform power is crushing the net" and has allowed people to "weaponise the web at scale" referring to the fake news crisis.[6] He also stated that the current web has been captured by corporations and that it has become a set of "walled digital gardens".

Another pioneer of the early days of the internet, Marleen Stikker also declares that the internet is broken. Being a thought leader on the digital society, she shares Berners-Lee's view that the promise of the internet has been wasted and perverted. She uses the analogy of green vs. brown fields, where green represents the virgin unspoiled earth and brown represents the divested and used up planet. The current online environment has become this brown field earth. An environment plagued by uncertainty about your own data, spying and hacking. It is a web where data black boxes, AI, and the illusion of objective data create serious dangers and disadvantages for users.[7]

In her 2019 book "Surveillance Capitalism", Shoshana Zuboff sounds the alarm bells that the power of big digital technology companies has become

---

[3] Maddux, C. D., & Johnson, D. L. (1997). The World Wide Web: History, Cultural Context, and a Manual for Developers of Educational Information-Based Web Sites. Educational Technology, 37(5), 5–12. http://www.jstor.org/stable/44428413

[4] The World Wide Web: History, Cultural Context, and a Manual for Developers of Educational Information-Based Web Sites on JSTOR

[5] Berners-Lee, T. (2018, 22 October). One Small Step for the Web. Inrupt. https://www.inrupt.com/one-small-step-for-the-web Accessed, 25/07/2022

[6] Lomas, N. (2019, 12 March). Marking 30 years of the web, Tim Berners-Lee calls for a joint fight against disinformation. TechCrunch. Marking 30 years of the web, Tim Berners-Lee calls for a joint fight against disinformation | TechCrunch. Accessed, 26/07/2022

[7] Stikker, M. (2020, July). Open The Black Box. https://waag.org/sites/waag/files/2020-07/Marleen%20Stikker-Essay-VNG%20Smart%20Society%20Cases%20Open%20the%20black%20box.pdf Accessed, 23/07/2022

a threat to our democracies, sovereignty, and ultimately to our humanity.[8] The extractive business model of "Big Other", as she calls these companies, turns human experience into profit, by feeding (advertising) engines of behavioural modification. Not only have humans been reduced to sensors, but they are also being actuated at a massive scale. This business model has for too long been left unchecked because the governments and law-makers have failed to understand the digital age and provide countervailing power in the form of legislation.

1.3 Promising Responses

The "broken state" of the internet, the slow take-off of smart cities with their low level of citizen engagement, is being remedied. Various EU legislations will curb the power of "big tech" on the internet, and personal data vaults may become the trusted tools that citizens need to engage.

1.3.1 European legislation

In Europe, 90% of its citizens believe that data should be uniformly protected across its member states.[9] To this end, the EU has created several institutions, rules, and laws, of which GDPR and the Data Act are particularly relevant to the use of personal data.

GDPR: The General Data Protection Regulation introduced in 2016 regulates the processing of personal data and its flows. It strengthens the fundamental rights of EU citizens by making clear the rules for private and public actors in the unified digital market. This single EU law also removes the current fragmented systems of different member states. National data protection authorities are hosted in each EU member state, to monitor GDPR adherence.

DGA and the Data Act: The goal of this legislation is to create an adaptable data ecosystem. The Data Governance Act (DGA) strengthens the single market's governance mechanism and establishes a framework to facilitate general and sector-specific data-sharing. The newly proposed Data Act addresses the actual rights on, access to, and use of data. These acts make it easier to harvest and use personal data. Particularly, by complementing the

[8] Zuboff, S. (2019). The Age of Surveillance Capitalism: The Fight for a Human Future at the New Frontier of Power. Public Affairs.

[9] The European Commission. (2022). Data protection in the EU. European Commission. https://ec.europa.eu/info/law/law-topic/data-protection/data-protection-eu_en Accessed, 27/07/2022

right to data portability of GDPR, making it easy to switch between providers or request that your data is given to your personal data-vault provider.[10]

DSA and the DMA: The Digital Markets Act and the Digital Services Act are two laws that limit the unbridled power of digital technology companies, particularly their "winner takes all" platform business models. As personal data will be shared through platforms, these laws may build the trust that citizens need to have in the systems and organisations behind these platforms.

1.3.2 Taking back control with data vaults

Another reaction to the broken promise is the development of means to give individuals control of their personal data. One of the originators of this idea is Tim Berners-Lee. He co-founded the start-up Inrupt, in order to allow governments and organisations to give control of data back to individuals, by means of solid personal data vaults (PDVs) and the associated platform.[11,12]

These PDVs are individually controlled safe data stores and key to privacy. The PDV separates the collection and storage of personal data streams from their dissemination. Instead of people directly exchanging their personal data streams with services, they use secure containers that only the user can fully access. With various forms of access control lists, the PDV permits the selective sharing of subsets of this information. This way, the data owner can actively participate in influencing data sharing decisions rather than depending on third parties. Ease of use and understandable data tools are key requirements.[13]

The European Union has also recognised the potential of this approach and has created an initiative to give citizens control over their personal data, namely the DataVaults initiative. The goal is to mobilise a movement that will advocate the removal of the barriers that currently prevent a citizen from sharing their data more broadly, whilst staying in control of the process.

---

[10] Considerati. (2021, 12 July). European Data Act & Data Governance Act: the kids with game changing potential. https://www.considerati.com/publications/eu-data-act-data-governance-act.html Accessed, 01/08/2022

[11] Solid. (z.d.). https://solidproject.org/origin Accessed, 01/08/2022

[12] Solanki, M. R. (2020). SOLID: A Web System to Restore the Control of Users' Personal Data. Advances in Intelligent Systems and Computing, 257–267. https://doi.org/10.1007/978-981-15-8289-9_24

[13] Mun, M., Hao, S., Mishra, N., Shilton, K., Burke, J., Estrin, D., Hansen, M., & Govindan, R. (2010). Personal data vaults. Proceedings of the 6th International COnference on - Co-NEXT '10. https://doi.org/10.1145/1921168.1921191

This will help the drive towards smart cities and galvanise the adoption of urban data platforms. The availability of personal data would expand as a result, stimulating the data economy and improving the state of the data ecosystems. All players can reuse data once individuals have given their approval.[14]

1.4 Think!

The internet did not pan out as the founder of the World Wide Web as early internet pioneers envisioned. Governments were late in pushing back against the domination by "big tech". Governments and companies alike find it hard to entice citizens to engage within the smart city. Yet, even with this *shaky track record*, we are enthusiastically paving the way for the next big, data driven, innovation: personal data vaults in the smart city.

Maybe this time, with the experiences of the smart city and the internet in mind, we can try to think a bit more before we succumb to our increasingly insatiable hunger for data. Far from being Luddites, we feel that taking a long view and assessing both the risks and opportunities of using personal data is no luxury. Berners-Lee had a good reason for wanting to build personal data vaults: to give control back to us. It was not necessarily to make it easy for smart cities to "hoard data", especially not personal data or for fuelling the data economy that wants to turn all pockets of unused data into gold.

Trying to think ahead is difficult but necessary, especially in these times of *profound change and great uncertainty*. In his 2020 new year's speech, Antonio Guterres of the United Nations described our times of uncertainty as the four horsemen of the apocalypse: geopolitical tensions, the climate crisis, global mistrust, and the dark side of technology.[15] According to the Edelman Trust Barometer,[16] distrust seems to have become society's default emotion. Media and the government are viewed as divisive forces, with less than half of the respondents trusting government leaders (42%) and journalists (46%).

---

[14] DataVaults. (2022). Smart Cities Marketplace – Citizen Control of Personal Data Initiative. DataVaults I EU. https://www.datavaults.eu/material/liaisons-relevant-links/citizen-control-of-personal-data-initiative-citizen-focus-action-cluster/#about-citizencontrol-initiative Accessed, 01/08/2022

[15] United Nations. (2020, 27 January). UN chief outlines solutions to defeat 'four horsemen' threatening our global future. UN News. https://news.un.org/en/story/2020/01/1055791 Accessed, 27/07/2022

[16] Edelman. (2022). 2022 EDELMAN TRUST BAROMETER. https://www.edelman.com/sites/g/files/aatuss191/files/2022-01/2022%20Edelman%20Trust%20Barometer%20FINAL_Jan25.pdf

Trust in government is lower than trust in companies in 23 out of the 28 global countries surveyed, and in all seven of the European countries surveyed, with social media companies as a noteworthy exception.

In these times of profound change, we must make an effort to curb the possible "dark side of technology". By using scenario thinking, we aim to increase awareness of the early warning signs of crossing over to the dark side. But first we take a closer look at what it means to have a personal data vault.

1.5 Personal Data Vaults Matter

The personal data vault might become the most important tool in our journey as a citizen. Balancing the benefits against the pervasiveness of this tool should be a conscious decision, subjected to continuous evaluation.

1.5.1 Capturing and influencing the citizen journey

In the digital age, individuals generate data from the cradle to the grave. Every aspect of our life, including our emotions, is being digitised and, if possible, quantified. It starts when we are born and stops when we die (see Figure 1.3). All this data becomes food for algorithms and machine learning. For instance, from our educational data and social background, one can predict the likelihood of study success. Study advisors can use this to intervene, hopefully with our best interest at heart. Our healthcare records

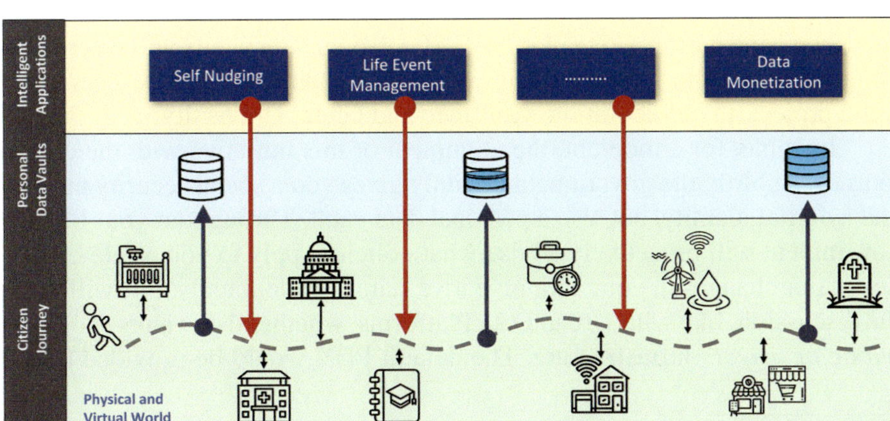

Figure 1.3 A citizen journey reveals points of sensing and actuation using personal data vaults.

in combination with our location data give public health institutions insights into the environmental and social impact on health. And as we all know by now, our browsing, shopping, and social media behavioural data are the "oil" that fuels the big tech advertising machines.

So, along our life journey, data is not only generated and harvested but also used to steer our journey into a particular direction. And this *steering or actuation* is happening at a massive scale, using all kinds of smart devices, in particular the smart phone. The idea behind EU legislation is that we know which data is being collected about us and that we have the right to summon this data and store this in our personal data vault. We can then decide who we want to share this data with and by whom we want to be steered during our journey. Can we?

1.5.2 Who will help us?

Can we expect citizens to be savvy enough to know who to share their data with, and under which conditions? Are we not on the path of increasing the *digital divide*? Or will there be benevolent parties, like specialty data trusts or stewards, that will help us to manage our data and, to some extent, thus our lives? Would we trust the government to this on our behalf?

The Flemish Data Utility Company[17] is a very interesting government initiative focusing on responsible and secure data sharing to increase citizens' trust in exchanging data. At the same time, the initiative aims to stimulate the Belgian economy by facilitating data discovery and interchange and by fostering collaboration. They hope to act as an impartial third party, a catalyst for creative projects, and an engine for societal and economic success. An "advanced and cutting-edge platform" of personal data vaults will be created. Flanders is the first government to use the data vaults of Tim Berners-Lee's Solid project, in collaboration with Imec and Ghent University.

Imagine, for a moment, the alignment of this initiative with the citizen journey. At birth, the government not only gives you a social security number and a digital identity, but also a personal data vault. Throughout your life, the government will proactively advise what policies apply to you at life-events, i.e., major changes in your administrative status. Undoubtedly, this will spark the discussion, like with Urban Data Platforms, whether data vaults should be *public or private* infrastructure. The default PDV would be provided by the

[17] Het Vlaams Datanutsbedrijf. (z.d.). The Flemish Data Utility Company. Digitaal Vlaanderen. https://www.vlaanderen.be/digitaal-vlaanderen/het-vlaams-datanutsbedrijf/the-flemish-data-utility-company Accessed, 01/08/2022

government, but the baby's parents are free to choose another provider. Why would citizens want to participate in this data sharing?

1.5.3 Personal benefits of the PDV

We would like to emphasise three benefits of personal data vaults: the protection against online manipulation, the enhancement of city services, and the possibility to change individual behaviour.

Lower online manipulation:
The core idea is that data vaults can be used to self-determine who uses your data and for what purpose so that you are aware that a particular service is tailored to your profile, which, when done right, is also a reflection of your needs. Most of us consent to the terms of use of apps without reading them and without really questioning how our data is being used. The research by Pew shows that about half of Facebook users are not comfortable with how they are categorised.[18] With a PDV, we still cannot control how we are categorised, but we can control what data is used to categorise us.

Better city services:
The Horizon 2020 project "Data Vaults" showcases a wide range of examples of how personal data can be used to improve city services, e.g., healthcare data sharing, smart home energy data sharing, and using personal data to improve services in the tourism industry. Data can be used at an aggregated and at an individual level. At the individual level, services can be personalised. But caution is advised. Personal data driven policies and services in the social domain have been criticised for leading to drive containment and control, rather than improve services.[19]

Cities find it hard to engage citizens. Interestingly, when asked about how citizens could be better engaged in the usage of an Urban Data Platform, in the RUGGEDISED UDP study, representatives from both the public and private sectors, whilst disagreeing on other aspects of citizen engagement, gave a high priority to "monetisation" of personal data (see Figure 1.4). Apparently in the mind of practitioners and policy-makers, citizens need to be incentivised to help improve city services.

---

[18] Hitlin, P. and Rainie, L.., Pew Research Center (2019), Facebook Algorithms and Personal Data. Facebook Algorithms and Personal Data | Pew Research Center

[19] Van Zoonen, L., (2020) Data governance and citizen participation in the digital welfare state. Data & Policy, vol.2 no. E10, pp. 1 – 16.

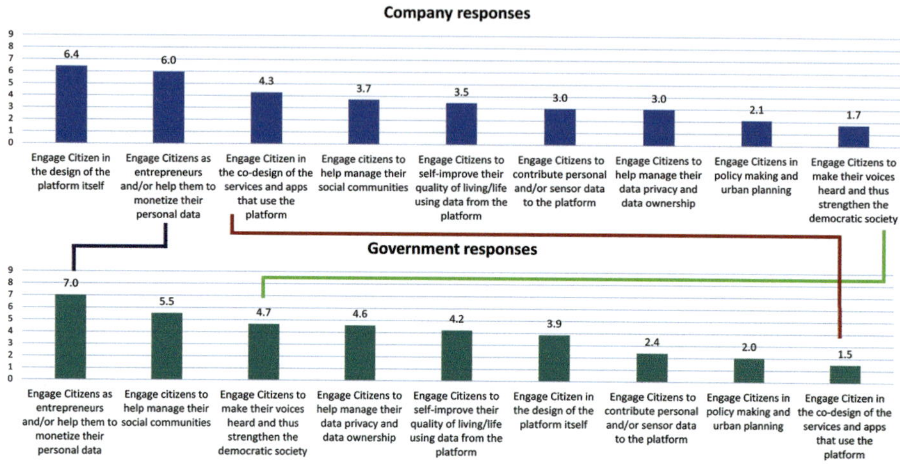

Figure 1.4 Reasons for companies and companies to engage citizens differ (source: RUGGEDISED study).

Impact on self-nudging:

With personal data comes the opportunity of self-nudging. Smart devices such as the smart phone are with us 24 × 7 and are the ultimate "nudging" device. As we have seen, along the customer journey, there are many potential points of intervention, and in the case of self-nudging, we decide what those points are. To this end, we need to have a smart personal digital assistant (PDA), which we must program to meet our needs and will help us get closer to an ideal virtuous life.

Do these benefits of the PDV outweigh its pervasiveness and potential intrusiveness? That is the question. Not getting this balance right may drive us toward dystopia, even when we have the best intentions of making the world a better place.

1.6 Utopia or Dystopia? A Scenario Analysis

The use of personal data will undoubtedly have both upsides and downsides. We want to consider the downside upfront. By taking a long view using the scenario thinking technique, we aim to understand how things could work out. The purpose of this forward looking thought experiment is to heighten the awareness of all the many pitfalls, as we pursue personal data utopia. The kind of pitfalls that we failed to envision in the case of the internet and the smart city. What could be the outcomes of infusing personal data into the smart city ecosystem? And what measures are needed to mitigate the risks?

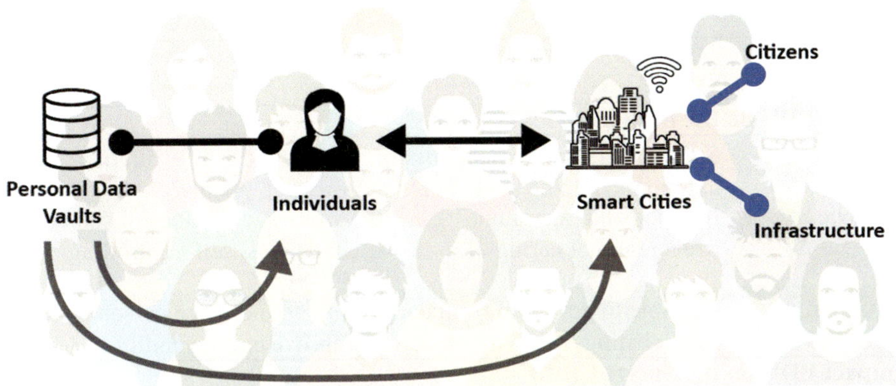

Figure 1.5 What is the impact of personal data on individuals and on the city?

1.6.1 Scenario drivers

Both the idea and technology of personal data sovereignty are in their infancy, and their impact is hard to predict. Therefore, we will paint a dark and a sunny picture for the impact on the individual and on the citizen. We ask two questions (see Figure 1.5):

1. What are the potential positive and negative impacts of personal data (vaults) on the individual?
2. What are the potential positive and negative impacts of personal data (vaults) on the city?

These two drivers will subsequently be used to construct four scenarios in the next section. Table 1.1 gives a high-level summary of these scenario drivers.

First driver: Impact on the individual:
The positive impact of personal data on the individual is *empowerment*. Data can be used to alter one's behaviour change for good, e.g., through healthy living, and prudent financial management. Service providers in both the public and private sectors can use personal data to personalise services leading to greater customer and citizen experiences. A well-functioning and trusted government may even proactively help citizens manage their life events, e.g., going to work, picking up children, etc. And last but not least, citizens can get value from their personal data, e.g., in health trials, or consumer research.

The negative impact of personal data is *subjugation*. People are reduced to their digital profiles and become cogs in a supposedly intelligent machine. If personal data is mindlessly shared with service providers, we will still be

Table 1.1 Highlights of both the negative and the positive impact on the individual and the city.

Impact PDV on the individual		
State of mind	Subjugated, manipulated loss of autonomy, feeling excluded	Empowered, free, full of potential, positive
Behaviour	Irresponsible, ignorant, irreverent, not caring	Virtuous, conscious of societal challenges, self-controlling (PDA)
Financial	Not able to get value from data, exacerbating the digital divide	Able to monetize personal data, improve employability
Impact PDV on the Smart City		
Economic	Too tightly coupled, not capable, no room for exploration	Efficient, well functioning, clean, good services, adaptive, vibrant
Political	Big brother-ish, intrusive, authoritarian	Engaged citizens, democratic
Collective, Community, Social	Segmented, profiling, un-equal, excluding, discriminating	Inclusive, acceptable level of inequality, coherent

manipulated, coaxed into echo chambers, and will risk losing our privacy; only this time, with our explicit consent. Surveillance capitalism will be fortified and we may end up in different digital castes depending on our digital savviness. We may even (temporarily) lose some of our autonomy, if governments deem it necessary that we render our personal data, all nicely wrapped into a single box; e.g., in case of emergencies or other reasons that serve the "greater good". In the Netherlands, reluctant psychologists and psychiatrists are mandated by healthcare authorities to share patient data.[20]

Second driver: Impact on the city (i.e., public and private actors in the city): The positive impact of personal data on the city will lead to a *Vibrant City*. This city is well run because of improved service provision based on aggregated personal data. For example, monitoring the movement of citizens using their GPS data enables better operational and tactical design of mobility infrastructure and services. The city is more efficient and resources become available to invest in the quality of life in the city. Improved social policies,

[20] Blijker, J. D. (2022b, July 21). Toezichthouder verzamelt ongevraagd info over 800.000 ggz-patiënten. Trouw. Accessed 8 August 2022, https://www.trouw.nl/binnenland/toezichthouder-verzamelt-ongevraagd-info-over-800-000-ggz-patienten~b68a0bef/?-referrer=https%3A%2F%2Fwww.google.com%2F

using citizen data, may lead to more engaged citizens and less inequality, which are good for democracy.

A city negatively impacted by personal data will become a *corrosive city*. Based on aggregated data and an efficiency mind set, the city becomes overly efficient, not leaving any room for exploration and diversity. Because not everyone can benefit equally from the use of personal data, we widen the digital divide, increase inequality, and lose sight on part of the population. People that do not fit into mould become outliers, are excluded, and may ultimately disengage. Bias of either institutions or the collective, manipulation, and abuse by "big brother on steroids" is among the biggest risks.

1.6.2 Four scenarios

If we combine the positive and negative impacts on cities and individuals, we get four scenarios as depicted in Figure 1.6. In Utopia, personal data is used for the betterment of the city as well as the citizens. In Dystopia, the government and companies will use our personal data to constrain us, while many individuals will not be able to leverage the data to their benefit and disengage.

The four scenarios can be described as follows. Please keep in mind that this is merely illustrative and no attempt is made to be academically precise. The purpose of this exercise is to ignite our collective imagination and create an *early warning system* for the abuse of personal data (vaults).

Unacceptable stagnation: Even though individual citizens thrive, e.g., by monetising their data, they will not accept the constraining living circumstances of the city. Excessive profiling will divide the city, physically and politically. The ecosystem will fail to develop good city services, and the political climate and tight government control squelch entrepreneurship.

Utopia: Both the city and the individual flourish from the use of personal data. People are empowered and engaged. There is private and individual initiative to tackle societal problems, e.g., climate change through innovation and behavioural change. Great city services attract people to the city and turn our cities into thriving hives of innovation.

Dystopia: This city much resembles Oceania in George Orwell's novel "1984". Through the use of personal data, both companies and the government have a firm grip on citizens. People share their data mindlessly, exposing themselves to manipulation. Ultimately, they feel exploited and will stop trying to live a virtuous life. In today's Dystopia, Orwell's "Big Brother" and Zuboff's "Big Other" are good friends.

Unsustainable subjugation: Government and companies are able to extract value from personal data. With personalised services, the government

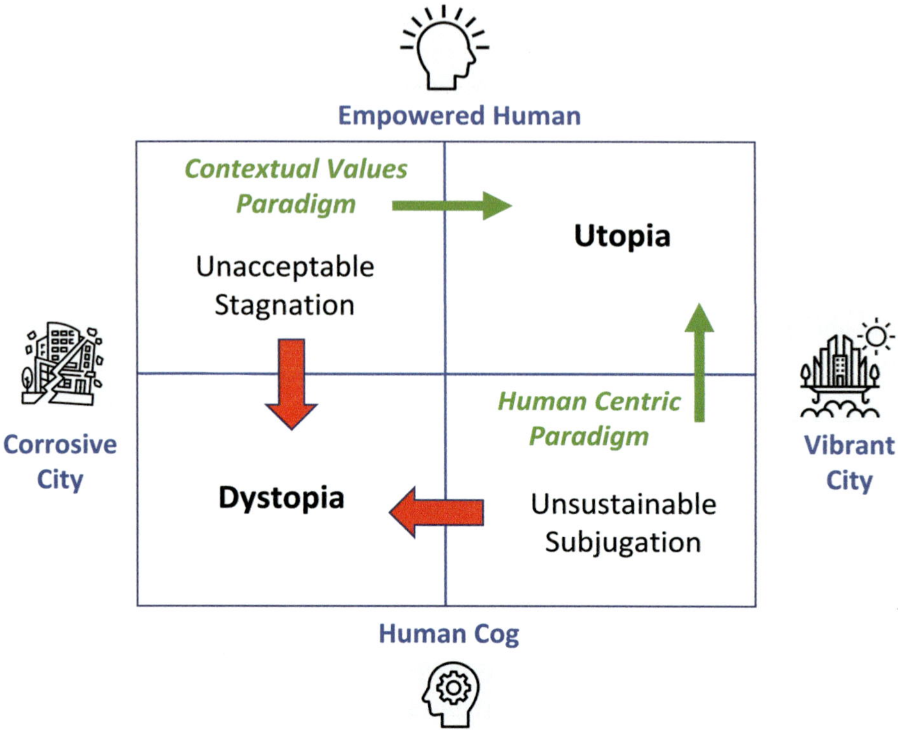

Figure 1.6 The future impact of personal data (vaults) in terms of four scenarios.

is even capable of engaging citizens into democratic processes. However, citizens feel objectified and are just "cogs in the machine". They will lack initiative and ultimately, without good and smart citizens, the well-organised and liveable city will deteriorate into decline.

1.6.3 Strategies to reach Utopia

The scenarios "unacceptable stagnation" and "unsustainable subjugation" are inherently unstable. It is very unlikely that empowered individuals will remain so in a corrosive city or that a city will remain efficient if the majority of its citizens feel alienated. Therefore, in our thought experiment, these scenarios either progress toward "utopia" or regress into "dystopia". From thermodynamics, we know that a system left to its own devices will deteriorate into chaos. So, to prevent dystopia from happening, we need to have *active strategies* to push these two scenarios towards "utopia". On a high level, we see two strategies (see Figure 1.6).

A *"human centric" paradigm* ensures that we have the citizen at the top of our mind when devising policies and designing our cities. A human-centric paradigm asks that we do not focus on the data economy only, i.e., seeing data as a means to generate economic value. Abandoning an economy-first paradigm is not easy but necessary. To get to Utopia, the twin transition should actually be a "triple transition", making the social transition as important as the energy and digital transitions.

With a *"contextual values" paradigm*, we expect the government and the rest of the city ecosystem to create an environment for individuals to thrive. Doing so requires us to move our attention away from the individual per se, to the context in which that individual can thrive. A context that abides by our European values. A context that allows for exploration and creativity, which stimulates diversity and by doing so creates a city where all individuals, today and tomorrow, can fulfil their full potential and play their part in solving today's challenges.

With these strategies in mind, we will not only focus on the nuts, bolts, and benefits, but we will also be mindful of possible adverse effects or blatant misuses of personal data. At least, that is our idea: preventing us all, unlike Icarus, from "flying too close to the sun".

1.7 Personal Data: "Fragile, Handle with Care"

Personal data vaults can be a force for good, both in terms of the empowerment of the individual and in terms of making our cities a better place to live. However, from the very outset, we should be mindful of what can go wrong when we deliberately capture, integrate, and unleash sensitive personal data into smart city ecosystems. So we do not repeat the mistakes of smart cities not being able to engage citizens or of the internet being hijacked by Big Tech. Thereto, at least three topics must be on our personal data and smart cities "radar".

Human dignity: At the base of human rights is human dignity. With rapid "datafication" of our lives, we run the risk of being reduced to our profiles, and as such, we become artefacts. Attacks on our inherent dignity include being objectified, treated as unequal, being invisible, and not having a voice.[21] It is not hard to imagine how the pervasive use of personal data might harm human dignity.

---

[21] Leidner, D.E., & Tona, O. (2021). The care theory of dignity amid personal data digitization. MISQ 45(1), pp 342–370.

Digital divide: There are two levels to the digital divide: one level is internet access and the other level is the skills people possess to benefit from digitalisation.[22] Not all citizens are equally tech-savvy or have the digital literacy to participate in the smart city. Research has shown that more in general socio-demographic characteristics also impact the use of digital technology.[23] By adding yet another digital instrument like the PDV, we are only exacerbating an already existing gap. Those that do not (know how to) share will feel "invisible", overwhelmed, and left behind. History teaches us that it is bad.

Algorithmic Bias: When we talk about using personal data, we are not talking about data only but also about the use of AI. The more data that is out there, the more AI will proliferate. All data will at one point in time be fed to algorithms and machine learning. When we rely too much on AI, or we can no longer explain how AI works, we run the risk of losing control. Bias, lack of transparency, and lack of accountability are a few of the issues that are widely recognised.[24]

The promise of smart cities with personal data is the promise of a human-centric smart city as described, e.g., by UN Habitat. It is the promise of a city with smart effective infrastructure as well as smart engaged citizens. A city where those that cannot cope with the speed, complexity and uncertainties of the digital age are helped, so they do not become prey for undemocratic forces. It is a city where the entire ecosystem, all actors in the quadruple helix, play their part in solving the grand challenges of our time. This is no small promise.

To fulfil this promise, we will need at least the following:

- the right privacy enhancing technology that respects human rights and ensures responsible AI;
- the right governance to implement the strategies that mitigate the "regression to dystopia";
- the right leadership, ethics, and values from citizens and institutions to obliterate the digital divide and safeguard human rights;

---

[22] Hargittai, E. (2002). Second-Level Digital Divide: Differences in People's Online Skills. First Monday, 7(4). https://doi.org/10.5210/fm.v7i4.942

[23] Shin, S. Y., Kim, D., & Chun, S. A. (2021). Digital Divide in Advanced Smart City Innovations. Sustainability, 13(7), 4076. https://doi.org/10.3390/su13074076

[24] Bird, E., Fox-Skelly, J., Jenner, N., Larbey, R., Weitkamp, E., & Winfield, A. (2020, March). The ethics of artificial intelligence: Issues and initiatives. European Union. https://doi.org/10.2861/6644

- the right research to fully understand digitalisation and the behavioural, economic, and political dynamics of our augmented humanity.

This is a lot to ask from ourselves as citizens and our institutions. It is fair to assume that we should not leave the use of personal data to the engineers and technocrats that have so far focused on the smart infrastructure part of the city. We need designers and practitioners who understand the opportunities and risks and who are capable of crafting and guarding a "path to Utopia". Because with the combination of personal data and other data, the active engagement of citizens willing to change their behaviour, and the current legal and financial stimuli from governments, the potential solutions for a better world are limited only by our imagination.

2

The Principal Projects Underpinning This Work

Ana García Garre[1], John Kaldis[2] and Elena Palmisano[3]

[1]MIWenergia, Spain
[2]Unisystems, Greece
[3]PIN, Prato

Abstract

This chapter describes the set of projects that we have concentrated upon to help map out the landscape and to draw out the potential benefits that can arise through combining a citizen's personal data with existing available data. It also draws attention as to why the European Commission decided to fund these projects, a major one being to support the emergence of data markets and the data economy, and, specifically, to set up and operate platforms for secure and controlled sharing of "closed data" (proprietary and/or personal data). These projects were also funded to address the necessary technical, organisational, legal, and commercial aspects of data sharing/brokerage/trading, whilst building on existing computing platforms.

2.1 Project Overviews

Below, we briefly describe the set of projects that we have concentrated upon to help map out the landscape and draw out the potential benefits that can arise through combining a citizen's personal data with existing available data. But it is worthwhile also drawing attention to the background for these projects in relation to the policies and the objectives that the European Commission have based their decision to fund these projects.

DOI: 10.1201/9781003399384-3

The focus for this book is around those projects that addressed the topic of "Supporting the emergence of data markets and the data economy".[1]

More specifically, the majority of the projects referred to, and from which the examples of the emerging technologies and the practices that the various demonstration cities have adopted to demonstrate these emerging technologies have come from, are the "innovation actions" for setting up and operating platforms for secure and controlled sharing of "closed data" (proprietary and/or personal data).

These projects were funded to address the necessary technical, organisational, legal, and commercial aspects of data sharing/brokerage/trading, whilst building on existing computing platforms.

With specific regard to the future use of citizen's personal data, the intention was that the emerging personal data platforms should be created to ensure compliance with a set of policy goals which included that:

- The personal data platforms shall ensure respect of prevailing legislation and allow data subjects and data owners to remain in control of their data and its subsequent use.
- Solutions should preserve utility for data analysis and allow for the management of privacy/utility trade-offs, metadata privacy, including query privacy.
- Solutions should also develop privacy metrics that are easy to understand for data subjects and contribute to the economic value of data by allowing privacy-preserving integration of independently developed data sources.
- Industrial data platforms shall enable and facilitate trusted and secure sharing and trading of proprietary/commercial data assets with automated and robust controls on compliance (including automated contracting) of legal rights and fair remuneration of data owners.

Other aims, if achieved, will also impact upon the growth of these personal data platforms.

- The call for proposals also addressed the fact that the lack of trusted and secure platforms and privacy-aware analytics methods for secure sharing of personal data and proprietary/commercial/industrial data

[1] "Supporting the emergence of data markets and the data economy - Programme," *H2020 - CORDIS European Commission*. https://cordis.europa.eu/programme/id/H2020_ICT-13-2018-2019 (accessed Jul. 22, 2022).

hampers the creation of a data market and data economy by limiting data sharing mostly to open data.

- It further intended to address the lack of ICT and data skills which were regarded as seriously limiting the capacity of Europe to respond to the digitisation challenge of industry.
- Specific attention needs to be put in involving SMEs and give them access to data and technology. IT standardisation faces new challenges as technologies converge and federated systems arise, creating new gaps in interoperability.

Collectively, they will stimulate the European data economy,[2] and of significance for this work, will in doing so, also stimulate the local data economy of a smart city, giving a stimulus in the form of releasing more valuable data for their use.

In addition to the core focus on personal data and the evolving data economy, we have drawn from other projects that have succeeded in bringing forward good examples of the use of data in a smart city. These have been drawn from a variety of sources including from the 18 smart cities and communities funded Lighthouse projects, which embrace 48 Lighthouse cities augmented by 72 "fellow cities" collaborating with them, to speed up replication. These Lighthouse cities are cities that are piloting and deploying the most advanced and innovative solutions.

Further projects highlighted have been funded to meet the identified societal challenges at a European level such as in health, demographic change, and wellbeing, where improved understanding of health and disease is demanding close linkage between fundamental, clinical, epidemiological, and socio-economic research. Effective sharing of data, standardised data processing, and the linkage of these data with large-scale cohort studies are also essential as are improving health information and better collection and use of health cohort and administrative data, with standardised data analysis and techniques.[3]

Other examples have come from the use of the European cloud infrastructure by public administrations in order to provide an infrastructure with data and analytical power for the public administration.[4]

---

[2] "Building a data economy - Shaping Europe's digital future." https://digital-strategy.ec.europa.eu/en/library/building-data-economy-brochure (accessed Jul. 22, 2022).

[3] "SOCIETAL CHALLENGES - Health, demographic change and well-being" *Programme H2020 - CORDIS European Commission*. https://cordis.europa.eu/programme/id/H2020-EU.3.1. (Accessed Jul. 22, 2022).

[4] "Pilot on using the European cloud infrastructure for public administrations - Programme," *H2020 - CORDIS European Commission*. https://cordis.europa.eu/programme/id/H2020_DT-GOVERNANCE-12-2019-2020 (accessed Jul. 22, 2022).

2.2 DataVaults

DataVaults[5] aims to deliver a framework and a platform that has personal data, coming from diverse sources in its centre and that defines secure, trusted, and privacy-preserving mechanisms allowing individuals to take ownership and control of their data and share them at will, through flexible data sharing and fair compensation schemes with other entities or for simply altruistic purposes.

In order to let the data providers decide what, how much, and in which manner they would like to share their data, we are using technologies that enable data asset control and management, trusted and secure sharing, and risk exposure awareness.

In order to guarantee the privacy and security of personal data during collection, sharing, and utilisation, the technologies embraced include TPM attestation/crypto, private and public ledgers, ABAC +ABE, layered anonymisation, and searchable encryption.

In order to attribute a fair share of the value data generated and to allow data seekers to extract intelligence, novel compensation schemes were developed along with blockchain-based data access and sharing accompanied by data analytics and intelligence generation.

The actors are the data owners (providers) on one hand and the data seekers (consumers) on the other.

Owners are the individuals who extract, collect in one place and securely store their personal data, share them, and take control of their usage. Seekers are the economic operators who explore extracts or metadata, request access, and perform analyses.

The outcome will be citizens getting back the control of personal data and an ever-growing eco-system for personal data sharing, thus having the potential to improve service delivery, increase citizen participation, and stimulate the local data economy.

2.3 KRAKEN

Based upon the self-sovereign identity paradigm, the project is providing:

- A decentralised user-centric approach on personal data sharing.
- A data marketplace, which will allow the sharing of personal data and its corresponding artificial intelligence/machine learning analysis, all while preserving privacy.

[5] "DataVaults - Empowering Secure Data Storage, Sharing and Monetisation." https://www.datavaults.eu/ (accessed Jul. 22, 2022).

- A set of different analytic techniques based on advanced crypto tools that will permit privacy-preserving data analysis.

KRAKEN (BroKeRage and MArKet platform for pErsoNal data)[6] is developing a trusted and secure personal data platform with state-of-the-art privacy aware analytics methods (with guarantees on metadata privacy, including query privacy).

Returning the control of personal data back to users:

The KRAKEN project aims to enable the sharing, brokerage, and trading of potentially sensitive personal data, by returning the control of this data to citizens (data providers) throughout the entire data lifecycle. In all the developments of KRAKEN, GDPR has been followed and assessment provided by the legal partner.

KRAKEN will standardise different IT solutions, thanks to featuring the (privacy-preserving) integration of independently obtained data sources from subjects consenting to different analyses. The project combines, interoperates, and extends the best results from two existing mature computing platforms developed within two H2020 actions: CREDENTIAL and MyHealthMyData.

In this context, the KRAKEN solution is piloting in two high-impact domains, health and education, and contributes to data spaces by providing tools and solutions that can preserve the privacy and assuring security, trustability, data integrity, and confidentiality, when data are shared between different stakeholders in a data space, and even between different data spaces.

2.4 Safe-DEED

Safe-DEED's vision[7] is a competitive Europe where individuals and companies are fully aware of the value of the data they possess and can feel safe to use it.

As privacy and trust remain key in the data sharing debate, privacy enhancing technologies will play a prominent role by 2025. Safe-DEED takes a highly interdisciplinary approach, bringing together partners from cryptography, data science, business model innovation, and the legal domain to focus on improving security technologies, improving trust, as well as diffusion of privacy enhancing technologies to keep up pace with global macro trends and the data economy, to enable the fastest possible growth.

Furthermore, as many companies have no data valuation process in place, Safe-DEED provides a set of tools to facilitate the assessment of

---

[6] "KRAKEN - broKeRage And marKEt platform for persoNal data." https://krakenh2020.eu/ (accessed Jul. 22, 2022).

[7] "Safe-DEED ." https://safe-deed.eu/ (accessed Jul. 22, 2022).

data value, thus incentivising data owners to make use of the scalable cryptographic protocols developed in Safe-DEED to create value for their companies and their clients.

The contribution that Safe-DEED has made has been related to how value is put on data and how business models in the data economy are evolving and in indicating what work needs to be done in the next phase. Chapter 13 elaborates on this and starts relating to the valuation of data within the context of the "story of data" in a smart city.

The general concept of Safe-DEED is a virtuous cycle of establishing value, providing the enabling technologies and creating the value thus promised, all within a solid legal and ethical framework.

1. Develop new methods addressing the problems of private set intersection for large datasets, and secure multiparty computation (MPC), reducing both communication (Mb/s) and runtime (s) by at least 50% against the existing state-of-the-art, while improving the expressiveness of the class of functions evaluated on the data.
2. Develop the first of its kind software as a service (SaaS) component for evaluating the potential of a company's data if collected and processed at scale.
3. Show the scalability of the methods in real-world scenarios involving personal data (from Telecom provider ForthNet) and confidential industry data (from semiconductor manufacturer Infineon).
4. Show the general applicability and sustainability of the methods for MPC and data valuation.
5. Develop pricing models and business models to exploit both the MPC services as well as the data valuation services.
6. Generate trust in data markets through sound legal and ethical foundations in the use-cases, project output, and the development of an educational module; cross-sectoral guidance on ethical/normative valuation and how to better align technology and economic incentives with the law and ethics.

2.5 DUET Project

DUET[8] advances an area's transformation through utilising a new innovative public sector digital twin which is a continuously learning digital copy of real-world assets, systems, and processes.

[8] "DUET Digital Twins." https://www.digitalurbantwins.com/about-1 (accessed Jul. 22, 2022).

DUET enables decision-makers from multiple sectors to co-create innovative solutions to complex urban challenges by drilling down into data through a shared, easy-to-use interface. Although DUET's solution is designed primarily for cities and public administrations, there is no reason why the application cannot be used by other industries. The underlying architecture is built with interoperability in mind. Users can upload any model so long as it complies with the framework requirements, after which they will be able to run simulations, analyse data, visualise insights, and much more.

Built to handle large datasets, both historic and real-time, the solution enables interrogation of anomalies and deviations before a disaster strikes. Thanks to the IoT connection, investigations can be done remotely from any place within a city and beyond; all that is needed is access to the DUET dashboard. Other benefits come from embedded machine learning and artificial intelligence functions, which allow the system to make predictions concerning, e.g., depreciation of road infrastructure or environmental footprint. We start by testing DUET in the smart city environment, with local administrations as main users (Flanders, Athens, and Pilsen). However, a fully developed solution has no restrictions when it comes to potential adopters. City halls, companies (big and small), universities, and emergency services all are using DUET because it is cheaper, more advanced, and user-friendly than competitor solutions. It helps organisations become more responsive, reacting rapidly to real-time events; policy decisions are faster and more effective, and relationships with citizens are improved.

In looking at advantages of a digital twin to a city, it is estimated that only 12% of a city's data is actually used in decision-making. It can change the way in which we see a city. It is valuable for experimentation to find policy solutions that work. It consumes data from across the city and uses high-powered computing to distil enormous amounts of data in order to deliver insights. 3D interfaces contribute to making policy implications easier to be understood by everyone.

2.6 InteropEHRate

The project sets out to put EHR in people's hands across Europe, having the slogan "Electronic health records made easy for patients to manage".[9]

The electronic health record (EHR) collects, systematises, and stores patient data in a digital format in order to improve healthcare systems. However, there is a low level of systems interoperability in Europe since

[9] "InteropEHRate." https://www.interopehrate.eu/ (accessed Jul. 22, 2022).

data are collected in different silos and managed under converging security and safety conditions. This creates legal hurdles in the availability of data. The EU-funded InteropEHRate project will reverse trends by ensuring that health data are available when and where needed. It will provide patients with full control in usage and routes of their medical information through device-to-device and peer-to-peer protocol standards. It will also outline a set of new protocols for secure and cross-border exchange of medical evidence.

Today, citizens moving across Europe have very limited control on their own health data, spread out in different silos. Legal constraints may prevent controllers of these silos from exchanging the managed data, even in an anonymised way, without the intervention of higher authorities. As a consequence, health data cannot be fully exploited for healthcare and research.

InteropEHRate aims to empower the citizen and unlock health data from local silos, using a bottom-up approach for EHR interoperability.

1. Mediated by the citizen: through the adoption of a D2D (device to device) standard, it, by exploiting edge computing and short-range wireless technologies, allows the citizens to import their own health data on personal smart devices, and exchange them, in a confidential way, also without the internet, with healthcare professionals and researchers, without the intervention of other authorities.

2. Authorised by the citizen: through peer-to-peer protocols for cross-border interoperability among EHRs and research apps, using decentralised authorisation mechanisms based on citizens' consent, to guarantee data accountability and provenance traceability, in compliance with patients' rights and GDPR.

3. Open and incremental: based on open specifications, connecting for-profit and non-profit data providers with different levels of interoperability, starting from a low level for secure exchange of unconverted data to a high level combining knowledge extraction and adaptive data integration, to translate data to a common HL7 FHIR[10] profile and into the natural language of the consumer.

4. A co-design approach and a specific governance model will manage human aspects related to ethics, laws, and technology evolution.

[10] https://www.hl7.org.uk/standards/hl7-standards/fhir/

User scenarios, presenting different security and interoperability requirements, will be validated by citizens and institutions belonging to six European countries.

Existing interoperability infrastructures will be exploited, including Connecting Europe Facility building blocks such as eIdentity.

2.7 RUGGEDISED

RUGGEDISED[11] is a smart city project funded under the European Union's Horizon 2020 research and innovation programme. It brings together three lighthouse cities, Rotterdam, Glasgow, and Umeå, and three follower cities, Brno, Gdansk, and Parma, to test, implement, and accelerate the smart city model across Europe in partnership with businesses and research centres. The three overall aims of RUGGEDISED are as follows:

1. improving the quality of life of the citizens, by offering the citizens a clean, safe, attractive, inclusive, and affordable living environment;
2. reducing the environmental impacts of activities, amongst others by achieving a significant reduction of CO_2 emissions, a major increase in the investment and usage of renewable energy sources and an increase in the deployment of electric vehicles;
3. creating a stimulating environment for sustainable economic development, by generating more sustainable jobs, stimulating community involvement in smart solutions (as consumers and as producers), and boosting start-ups and existing companies to exploit the opportunities of the green digital economy and Internet of Things.

With RUGGEDISED, there is no simple application or topic under scrutiny. Instead, in addition to the focus for our interest in this project, the progress in deploying urban data platforms, there is a wide array of areas where work is progressing. RUGGEDISED will demonstrate in total 32 innovative and integrated smart solutions in the cross-section of energy, transport, and ICT.

These include the following:

- smart open data decision-making platforms;
- 3D city operations model;

[11] "RUGGEDISED - Smart city lighthouse project." https://ruggedised.eu/smart-solutions/smart-solutions-overview/ (accessed Jul. 22, 2022).

- smart waste management;
- smart thermal grids;
- thermal energy storage;
- smart electricity grids and eMobility;
- energy data management;
- end-user involvement in intelligent building controls;
- long-range wireless networks.

2.8 DataPorts

It is a data platform for the connection of cognitive ports.

DataPorts[12] will provide a data platform in which transportation and logistics companies around a seaport will be able to manage data like any other company asset, in order to create the basis to offer cognitive services.

Nowadays, only 3% of container terminals are automatised. However, the future of the port industry points towards smart ports since it is the only way to overcome the challenges and demands that arise in the sector, optimising port operations, enhancing the supply chain of operators and carriers, and reducing the emissions and waste.

Nevertheless, the maritime port infrastructure is quite complex. On a site, a large number of agents interfere in each port operation (retailer, freight forwarder, carrier, consignee, port authority, etc.); thus, stagnant silos of information are produced and the real potential of the data cannot be obtained. From this need appears the DataPorts project, a data platform for the cognitive ports of the future.

Valencia and Thessaloniki are the pilot sites, and in relation to the smart cities agenda, there is interest in dealing with the logistics in the port city both for freight and tourists arriving by sea.

2.9 EUHubs4Data

Most of Europe's SMEs lag behind in data-driven innovation. To tackle this problem, the EU-funded EUHubs4Data project[13] will build a European fed-

[12] "DataPorts - A Data Platform for the Connection of Cognitive Ports." https://www.dataports-project.eu/ (accessed Jul. 22, 2022).

[13] "EUHubs4Data - European Federation Of Data Driven Innovation Hubs." https://euhubs4data.eu/ (accessed Jul. 22, 2022).

eration of data innovation hubs based on existing key players in this area and connecting with data incubators and platforms, SME networks, artificial intelligence communities, skills and training organisations, and open data repositories.

A European catalogue of data sources and federated data-driven services and solutions will be made accessible to European SMEs, start-ups, and web entrepreneurs through the data innovation hubs. Cross-border and cross-sector data-driven experimentation will be facilitated through data-sharing, as well as data and service interoperability, becoming a reference instrument for growth in a global data economy and contributing to the creation of common European data spaces.

EUHubs4Data provides an integrated ecosystem aimed to stimulate greater participation of European SMEs and start-ups in the data economy. By providing easy, cross-border access to datasets, facilitating data sharing and assisting them with the skills, tools, and support, new innovative data-driven solutions and business models that are aligned with users' needs will be developed, thereby improving their digital competitiveness as well as the satisfaction of end-users.

2.10 i3-MARKET

The i3-MARKET project[14] addresses the growing demand for a single European data market economy by innovating marketplace platforms, demonstrating with industrial implementations that the data economy growth is possible. The i3-MARKET aims at providing technologies for trustworthy (secure and reliable), data-driven collaboration and federation of existing and new future marketplace platforms. Special attention is given to industrial data, particularly sensitive commercial data assets from both SMEs and large industrial corporations.

The i3-MARKET aims to overcome current data marketplace integration obstacles by developing the lacking building blocks (in the form of a software framework or toolkit) for data providers and consumers. By doing so, it enables the creation of a more trusted European data market economy.

The i3-MARKET v2 Backplane software is now publicly available to the community. The v2 Release is built based on industrial stakeholder's feedback, consultation events with SMEs, developers, industry stakeholders,

[14] "i3-Market." https://www.i3-market.eu/about/#project (accessed Jul. 22, 2022).

and collecting requirements and demands from developer communities and entrepreneurs.

The i3-MARKET project contributes towards the growth of a single European data market economy by enabling secure and privacy-preserving data sharing across data spaces and marketplaces.

It is well known that despite various research and innovation attempts working on big data management and sharing, there is no broadly accepted trusted and secure solution for federation of data marketplaces. The i3-MARKET is addressing this gap by developing lacking technologies and solutions for a trusted (secure, self-governing, consensus-based, and auditable), interoperable (semantic-driven) and decentralised (scalable) infrastructure, the i3-MARKET Backplane, that enables federation via interoperability of the existing and future emerging data spaces and marketplaces.

2.11 AURORAL

It is an architecture for unified regional and open digital ecosystems for smart communities and wider rural areas large-scale application.

AURORAL[15] focuses on increasing connectivity and delivering a digital environment of smart object interoperable service platforms able to trigger dynamic rural ecosystems of innovation chains, applications, and services. Thus, AURORAL contributes to increase economic growth and create jobs in rural areas and to tackle significant societal challenges and contributes to overcoming digital divide between rural and urban areas and to developing the potential offered by increased connectivity and digitisation of rural areas. AURORAL digital environment is demonstrated by cost-efficient and flexible cross-domain applications through large-scale pilots in five European regions. It builds on an open, API-based, interoperable, and federated Internet of Things (IoT) architecture and includes a reference implementation supporting flexible integration of heterogeneous services, bridging the interoperability gap of the smart object platforms and creating markets for services in rural areas.

H2020-AURORAL was launched in January 2021 aiming to contribute to increase economic growth in rural areas and to tackle significant societal challenges.

[15] "AURORAL." https://www.auroral.eu/#/ (accessed Jul. 22, 2022).

H2020-AURORAL is building a new concept, the smart communities, a new European paradigm for sustainable development.

Those are areas based on the use of innovative solutions to improve their resilience, building on local strengths and opportunities.

The main objectives are as follows:

- connect and share data collected locally through a secure and privacy-preserving framework;
- engage external technology and application providers in exploiting their data by offering advanced horizontal services to process and create value out of these data;
- participate in new dynamic online marketplaces as commodities services and online platform operators;
- implement an interoperable way and based on open application interfaces based on open standards.

Vertical services and tools:

AURORAL reference architecture aims to facilitate the integration of different vertical tools and service from a variety of rural domains. Some indicative areas are:

- digitalised energy;
- digitalised mobility;
- digitalised farming;
- digitalised tourism;
- digitalised health services.

2.12 REPLICATE

REPLICATE – REnaissance of PLaces with Innovative Citizenship And TEchnologies,[16] is a European research and development project that aims to deploy energy efficiency, mobility, and ICT solutions in city districts.

"Our vision is to increase the quality of life for citizens across Europe by demonstrating the impact of innovative technologies used to co-create

[16] "REPLICATE Project EU - REnaissance of PLaces with Innovative Citizenship And TEchnology." https://replicate-project.eu/ (accessed Jul. 22, 2022).

smart city services with citizens, and prove the optimal process for replicating successes within and across cities".

There are three lead cities (called smart city "Lighthouses"), which include San Sebastian in Spain, Florence in Italy, and Bristol in Great Britain. There are also a number of other "follower" cities that will look into replicating interventions in their cities including Essen (Germany), Nilüfer (Turkey), and Lausanne (Switzerland). In addition to this, there are also a number of "observer" cities, such as Guangzhou (China) and Bogota (Colombia), as well as international networks who will take part in learning and dissemination of results.

The REPLICATE team is working to accelerate the deployment of innovative technologies, organisational and economic solutions to significantly increase resource and energy efficiency, improve the sustainability of urban transport, and drastically reduce greenhouse gas emissions in urban areas. Therefore, the project aims to enhance the transition process to a smart city in three areas: energy efficiency, sustainable mobility, and ICT infrastructures.

Energy efficiency – saving energy consumption:

- up to 56% in relation to existing situation in building retrofitting;
- up to 35% in district heating.

Sustainable mobility – integrating sustainable EVs, recharging systems, and information mobility system.

Integrated ICT infrastructures:

Developing new sustainable and cost-effective services to citizens providing integrated infrastructures that improve efficiencies in the use of local public resources and the delivery of public services:

- new ICT model based on FI-WARE and open data management;
- new intelligent lighting system based on new LED technology;
- high-speed mobile wireless network based on post WIMAX technology.

2.13 PIMCity

Personal information management systems (PIMS) [17] aim to give back users control over their data, while creating transparency in the market. PIMCity

[17] "PIMCity - Building the Next Generation Personal Data Platforms." https://www.pimcity-h2020.eu/ (accessed Jul. 22, 2022).

offers tools to enable businesses to mature and reach sizeable user bases. The PIMCity approach is both ambitious and revolutionary.

Personal data are increasingly necessary for the development of products and services that must combine their use with respect to the privacy and intimacy of citizens. For there to be confidence in the personal data ecosystem, it is necessary, in addition to an adequate regulatory framework, to have tools that allow citizens and companies to make an ethical use of them without renouncing the development of new business opportunities. The EU-funded PIMCity project aims to increase transparency and provide users with control over their data.

Mission: Our mission is to ensure that citizens, companies, and organisations are informed and can make respectful and ethical use of personal data. The human-centric paradigm is aimed at a fair, sustainable, and prosperous digital society, where the sharing of personal data is based on trust as well as balanced and fair relationship between individuals, business, and organisations. We need, besides being informed, that all the multi-stake holders are proactive and that they get involved in this process that is in continuous evolution.

What we do: We select, classify, and assess information of interest, in terms of privacy and personal data management, for citizens, companies, and organisations. We search and test tools and novel mechanisms to increase users' awareness. We are looking for materials that facilitate the work of evangelisation in order to achieve a well-informed society, whilst promoting the use of the tools developed in the PIMCITY project.

2.14 smashHIT

The smashHIT project[18] is designed to solve the problem of consumer consent and data security in the connected car and in the smart cities environments. SmashHIT is formed by a consortium of nine organisations drawn from analytics, data security, car manufacturing, smart city infrastructure, and academia working collaboratively to deliver the benefits of shared connectivity to millions of consumers.

Enabling millions of EU consumers to benefit:

- The objective of smashHIT is to assure trusted and secure sharing of data streams from both personal and industrial platforms, needed to

[18] "Smash Hit - Solving Consumer Consent & Data Security for Connected Car and Smart City." https://smashHIT.eu/ (accessed Jul. 22, 2022).

build sectorial and cross-sectorial services, by establishing a framework for processing of data owner consent and legal rules and effective contracting, as well as joint security and privacy-preserving mechanisms.

- smashHIT aims to overcome obstacles in the rapidly growing data economy which is characterised by heterogeneous technical designs and proprietary implementations, locking business opportunities due to the inconsistent consent and legal rules among different data-sharing platforms actors and operators.
- A vision, strategy, and future ready framework.
- Diversity and inclusion are integral to smashHIT's vision, strategy, and business success. We recognise that leadership in today's global marketplace requires that we create environments where the best and brightest diverse minds – employees with varied perspectives, skills, and experiences – work together to meet consumer demands. The collaboration of cultures, ideas, and different perspectives is an organisational asset and brings forth greater creativity and innovation.
- The framework will provide methods and tools, such as smart data dispatcher, to assure common consent over data shared using semantic models of consent and legal rules.
- The new tools include traceability of use of data, data fingerprinting, and automatic contracting among the data owners, data providers, service providers, and volumes on data streaming from the usage of mass products with cyber–physical features (e.g., vehicles).

These data streams offer new opportunities to build innovative services, but their combination with other personal and industrial data is subject to complex ownership and consent aspects, as the data streaming from these products belong to persons or organisations who are owners or users of the products.

The overall concept can be found on the project website.[19]

2.15 PolicyCloud

PolicyCloud aims to make data-driven policy management a reality across Europe.

[19] "Smart Dispatcher for Secure and Controlled sharing of Distributed Personal and Industrial Data," 2021. Accessed: Jul. 22, 2022. [Online]. Available: https://www.smashHIT.eu/wp-content/uploads/2021/03/smashHIT_D1.3_Public_Innovation_Concept_v100.pdf (accessed Jul. 22, 2022).

The PolicyCloud project[20] will harness the potential of digitisation, big data, and cloud technologies to improve the modelling, creation, and implementation of public and business policy. PolicyCloud will deliver a unique, integrated environment of curated datasets, and data management, manipulation, and analysis tools. The project will address the full lifecycle of policy management using the data analysis capabilities of the European Cloud Initiative.

During the course of the project, our expertise will be practically demonstrated in four thematically distinct pilot use-cases to be run in diverse European cities, positively impacting both economic growth and the lives of the citizens concerned.

The solutions and tools developed by PolicyCloud will eventually become available from the European Open Science Cloud (EOSC), as public cloud services. As explained by the EC President von der Leyen at the 50th annual meeting of the World Economic Forum in Davos, the EOSC is a core element of the European Commission's vision to make Europe a global leader in data management.[21]

By 2023, PolicyCloud will roll out a series of novel data-driven policy management solutions set to drastically advance policy-making and benefit a wide variety of stakeholders.

Policy-makers at European, national, and regional levels, public administrations, non-governmental organisations, and standardisation bodies will be provided with the ability to make efficient and effective policy decisions through access to:

- cleaned, refined, structured, and trustworthy datasets emerging from the pilot use cases;
- analytical tools to enhance the predictive power of data;
- scenario simulations to model and evaluate policy impacts.

Academic institutions, research centres, individual researchers, and big data experts will be able to achieve better quality research outcomes through access to:

- solutions and policy-making services available through EOSC;
- previous project results upon which to build further.

[20] "Policy Cloud." https://policycloud.eu/ (accessed Jul. 22, 2022).

[21] "Keynote by President von der Leyen at World Economic Forum." https://ec.europa.eu/commission/presscorner/detail/en/SPEECH_20_102 (accessed Jul. 22, 2022).

In industry, big data, cloud, and solutions, providers will experience improved efficiencies and attract new business opportunities through access to:

- novel data management and analysis solutions;
- tools for cleaning and refining data;
- via, pilots, improvements in the quality of certain governmental datasets;
- the data marketplace as a shop window via which to offer new datasets.

Citizens resident in range of the pilot use-cases or impacted by future PolicyCloud adoption will enjoy improved quality of life through:

- opportunities to participate in policy creation through the iterative design and implementation methodology adopted;
- continuous improvement in policies through an iterative design approach which leverages newly available data on an ongoing basis and optimises policies simultaneously across multiple sectors.

Increasingly relevant and useful policies designed around homogeneous segments rather than entire populations.

2.16 IRIS: Co-creating Smart and Sustainable Cities

IRIS Smart Cities[22] is a collective of seven European cities working to meet the challenge of making the transition to smart and sustainable and has been developed around three lighthouse cities – Utrecht (Netherlands), Nice (France), and Gothenburg (Sweden), with follower cities. In a rapidly changing environment, they are tackling the need for new strategies that help cities to smartly integrate technological solutions, accompanied by the fact that cities can act as large-scale demonstrators of integrated solutions, whilst wanting to contribute to the socially inclusive energy and mobility transition.

The overall concept of IRIS is the transition strategy comprising five "tracks" that together provide a universal yet versatile framework to address both common and district-specific challenges. Within these five tracks, IRIS demonstrates a set of integrated solutions built on top of both mature and innovative technologies. The integrated solutions are defined on the basis of a common-shared know-how interchange among the lighthouse and follower cities, with a planned replication from the outset of the project.

[22] https://irissmartcities.eu/ NOTE, not to confuse, we later refer to a more technical project, also named IRIS, with Barcelona, Helsinki, and Tallinn as the cities involved.

The tracks under scrutiny are:

1. renewable and energy positive districts;
2. flexible energy management and storage;
3. intelligent mobility solutions;
4. digital transformation and services;
5. citizen engagement and co-creation.

Track 4 is of particular interest with its city innovation platform with open, standards-based application program interfaces (APIs) provides meaningful data and information services for households, municipality, and other stakeholders, allowing for a data market with new business models. It embraces:

- urban monitoring;
- city management and planning;
- mobility services;
- energy management.

For the purpose of this book, the interest also lies in the business models under consideration. For mainstream sustainability, developing and delivering green infrastructure and services needs to make financial sense. With this in mind, IRIS has aimed to put bankable business models for proposed integrated solutions into practice – helping European cities and regions reduce technical and financial risks and attract investors.

2.17 SmartEnCity

Cities play a key role in fighting climate change. Energy demand and CO_2 emissions are particularly high in urban areas. At the same time, urban density allows more alternatives for energy-efficient housing, eco-friendly transport, and service provision. SmartEnCity's[23] vision is to create smart zero carbon cities that are more sustainable and inclusive, improve citizens' quality of life, create jobs and wealth, and offer equal growth opportunities.

SmartEnCity aims to develop a systemic approach for transforming European cities into sustainable, smart, and resource-efficient urban

[23] "SmartEnCity.eu." https://smartencity.eu/ (accessed Jul. 22, 2022).

environments in Europe. We aim to develop strategies that can be replicated throughout Europe in order to:

- reduce energy demand;
- maximise renewable energy supply.

Activities include retrofitting in buildings, integrating infrastructures, developing sustainable mobility, and the intelligent use of information and communication technologies.

The SmartEnCity concept was defined, planned, and implemented in the three Lighthouse demonstrators Vitoria-Gasteiz (Spain), Tartu (Estonia), and Sonderborg (Denmark). The process will be replicated in the two follower cities of Lecce (Italy) and Asenovgrad (Bulgaria). We will encourage a SmartEnCity network of further cities interested replicating the approach elsewhere in Europe.

In addition to the core set of projects, we also need to refer to two wider initiatives contributing to the evolution of the European data economy: MyData and Solid.

2.18 The MyData Global Initiative

MyData Global[24] believes that the future belongs to companies that provide human-centric solutions to privacy and control of data. "The human-centric approach to data is aimed at a fair, sustainable, and prosperous digital society. In such a society, people get value from their data and set the agenda on how it is used. And for organisations, the ethical use of data is always the most attractive option".

MyData Global is an award-winning, international non-profit organisation which has over 100 organisation members and close to 400 individual members from over 40 countries on six continents. It helps organisations to build human-centric solutions and services and collaborates with local, national, and international stakeholders to advance ethical use of personal data.

The MyData community includes thousands of experts who meet regularly at the flagship event, which is thc international MyData Conference, which has been taking place annually since 2016.

MyData Global's purpose is to empower individuals by improving their right to self-determination regarding their personal data. We bring together professionals and organisations from business, legal, technology, and society

[24] "MyData." https://www.mydata.org/ (accessed Jul. 22, 2022).

perspectives to accelerate transformation for ethical and human-centric data sharing and use. The mission and values are described in the MyData Declaration which entrepreneurs, activists, academics, listed corporations, public agencies, and developers are encouraged to sign in order to help make the use of human-centric data a reality.

The MyData Declaration provides a common goal: to empower individuals with their personal data and acts as a basis for restoring trust in data-based business and for establishing a balanced and fair relationship between individuals and organisations.

2.19 The SOLID Initiative

Solid[25] is a specification that lets people store their data securely in decentralised data stores called Pods. Pods are like secure personal web servers for data.

- Any kind of information can be stored in a Solid Pod.
- Access to the data can be controlled in the Pod. Deciding what data to share and with whom (be it individuals, organisations, and/or applications) is down to the individual.
- Access can be revoked at any time.
- To store and access data in the Pod, applications use standard, open, and interoperable data formats and protocols.

A Solid server hosts one or more Solid Pods. Pods are where you store your data:

- each Pod is fully controlled by the Pod owner;
- each Pod's data and access rules are fully distinct from those of other Pods.

Pods come from a Pod Provider or can be self-hosted.

It is possible to have multiple Pods. They can be hosted by the same Pod Provider or by different providers or be self-hosted or any combination thereof. The number of Pods possessed as well as which Solid server or servers you use is effectively transparent to the applications and services that you use. This is because, in the Solid ecosystem, data is linked through your

[25] "Solid: Your data, your choice." https://solidproject.org/ (accessed Jul. 22, 2022).

identity and not through the specifics of your Pod. This is true for your own data as well as for data that others have shared with you.

Solid has the ability to store data in a way that promotes interoperability. Specifically, Solid supports storing Linked Data. Structuring data as Linked Data means that different applications can work with the same data.

With Solid's authentication and authorisation systems, the individual determines which people and applications can access their data. They grant or revoke access to any slice of your data as needed. Consequently, the individual can do more with their data, because the applications used can be granted access to a wider and more diverse set of information.

Share data with others and others sharing data with you is possible. This creates rich and collaborative experiences across a combination of both personal and shared data.

Solid applications store and access data in Pods using the Solid protocol.

Within the interoperable Solid ecosystem, different applications can access the same data instead of requiring separate data silos specifically for the applications.

3

Best Practice in the General Use of Data in a City

Annina Autero[1], Michail Bourmpos[2], Iván Carrillo Quero[3], Emanuele Geri[4], Hervé Groléas[5], Alisa Jashari[6], Christina N. Karaberi[7], María Jose Lopez Osa[8], Sari Maenpaa[9], Esben Pejstrup[10], Elena Palmisano[11], Fabio Pazzaglia[12] and Minna Sapyska[13]

[1]SURE Project, Tampere, Finland
[2]Piraeus, Greece
[3]ATOS, Spain,
[4]Information Systems Directorate, City of Florence, Italy
[5]Chief Digital and Information Officer, Grand Lyon, France
[6]Tampere, Finland
[7]e-trikala – Department of Research and Communication, Trikala, Greece
[8]Tecnalia, Spain
[9]Insta, Finland
[10]ICLEI, Germany
[11]PIN, Prato, Italy
[12]Information Systems Directorate, City of Florence, Italy
[13]Tampere, Finland
Email: [1]anniina.autero@tampere.fi; [2]mmpour@piraeus.gov.gr; [3]ivan.carrillo@atos.net; [8]MariaJose.Lopez@tecnalia.com; [9]sari.maenpaa@insta.fi; [10]esben.pejstrup@iclei.org; [11]elena.palmisano@gmail.com; [13]minna.sapyska@tampere.fi

Abstract

This chapter gives a "bird's eye view" of a variety of examples of how some of the "EU CITIES MISSION's 100 climate neutral and smart cities by 2030" are already using data from a wide variety of sources. They collectively will

DOI: 10.1201/9781003399384-4

bring a wide range of experiences to ensure that the identified good practices in data usage might be shared and built upon, in the drive to make data work to drive forward the climate-neutral agenda.[1]

The examples come from a wide range of application areas and reflect the variety of the emerging data spaces, with the majority coming from the "100 climate-neutral cities". Whilst useful in its own right to be able to see the variety of uses data is being put to in a city, to stimulate ideas and potential replication elsewhere, the main purpose is for it to be used as a building block. The following chapter will concentrate on how cities have used personal data in the process of running their city. But in addition to providing these examples, we will also try to indicate how, by adding personal data to the data currently being used in the cases described in this chapter, added value might be demonstrated.

3.1 Flanders, Belgium

Flanders can be seen as a network of interrelated cities forming together with Brussels a large metropolis area, along with cities such as Antwerp, Bruges, Leuven, and Ghent. Flanders alone can be considered as one single, smart region of 6.7 million people. The smart Flanders network brings the region together to create the necessary scale for smart city solution and data providers. It is anticipated that a Flanders digital twin will play a crucial role to open and democratise available smart city data to citizens, companies, and service providers and to use for co-created policy. Flanders started from its spatial data infrastructure and early digital twin prototyping.

The overall goals for taking a digital twin approach are to:

- create a smart region where all players can access available services and data;
- support cross-silo cooperation between sectors;
- involve citizens and companies active in policy-making processes to improve the quality of decision-making and acceptance of the outcomes;
- set up transferable services and data standards to maximise efficiency and open the market.

[1] "100 Climate-neutral Cities by 2030 – by and for the Citizens ," *European Commission.* https://ec.europa.eu/info/publications/100-climate-neutral-cities-2030-and-citizens_en (accessed Jul. 25, 2022).

In terms of the specific policy goals which the Flanders DUET Project Digital Twin[2] will concentrate on will be: "The design of new measures and the implementation of actions and evaluation of the success of actions foreseen in the Flanders Regional Mobility Plan, and in the Flanders Environment Plan which both aim for smoother mobility through actions that are kinder to the environment and which reduce the impact on human health". These regional plans are translated to specific measures on the local level that can be part of a DUET Digital Twin simulation.

"Currently, Flanders is a European hot spot for air pollution. Alternatives to car transport have been put in place to increase the daily level of physical activity among the population and reduce air pollution and global warming. To evaluate the impact of existing measures and help create new ones, Flanders will test and implement the Digital Twin to ingest and use vast amounts of data to help the city and its stakeholders explore correlations between the mobility, health and air data".

A practical example can be seen in the simulation of what might occur if a specific bridge were to be closed in GHENT – what would be the implications on the traffic and the feelings of the local population were this to happen – and hence what would be the effect on achieving the policy goal of reducing air pollution.

Flanders is also experimenting with SOLID.[3]

3.2 Pilsen, Czech Republic

Pilsen with a population of over 165,000 as such is a good example of medium sized European city. It has its industrial activity and acts as a centre for retail and entertainment whilst acting as an important hub for commuters and attracting tourists and visitors. In keeping with other similar sized cities, Pilsen faces significant challenges in the process of designing how the city will progress and in dealing with transport planning. In moving towards being a smart city, it is developing and implementing solutions in a variety of areas and these include mobility, security, business support, environment, ICT, and public participation. Pilsen has a wide range of datasets, GIS, and traffic modelling tools but not from an integrated solution.

[2] "Flanders Twin - DUET: Digital Urban European Twins." https://www.digitalurbantwins.com/flanderstwin (accessed Jul. 25, 2022).

[3] "Solid: Your data, your choice." https://solidproject.org/ (accessed Jul. 22, 2022).

Pilsen's motive for joining the DUET project[4] is similar to that of Athens. Pilsen is faced with a lack of complex and data-based planning across different sectors and policy areas that influence each other, hence the adoption of the digital twin approach. Its initial focus in creating its digital twin will be the interrelationship between transport and noise pollution in a 3D environment. Noise pollution in the city environment is influenced by many factors. A significant factor is the local road transport and especially the traffic volumes that pass by the built-up environment, the types of vehicles, and traffic conditions such as speed limits. Whereas, the other factors that play a crucial role in the city ecosystem are the overall urban design, urban morphology, land use, street distribution, street environment, and green infrastructure.

In relation to the city's policy goals, a driving force behind using digital twins is to help achieve the city's overall goals and, in particular, the issue of reducing noise pollution.

Pilsen is demonstrating the digital twin concept across transport and mobility, urban planning, and environment and wellbeing by focusing on two related policy areas:

- Environmental Noise Directive (2002/49/EC) which recognises the need to protect quiet areas in cities and towns as sites of value to the local community. Plants and specific land use can play a role in this by softening the urban environment and reducing noise.
- The development of Pilsen's sustainable mobility plan. Stress will be made on the interrelation of these policy areas, breaking the silo-based traditional approaches in decision-making.

3.3 Camden, London, United Kingdom

Open data policies for citizens:

The Camden Plan is the council's response to the Camden 2025 vision. This is our plan for how we, as an organisation, will play our part to achieve the ambitions set out in Camden 2025. It reaffirms our values and ambitions and shows how we will work with our communities to take forward our shared priorities over the next four years, between 2018 and 2022. This means working together to make sure that Camden stays an amazing place to live, work, and grow up, where everyone can lead happy, healthy, and fulfilling lives.

---

[4] "Pilsen Twin - DUET: Digital Urban European Twins." https://www.digitalurbantwins.com/pilsen-twin (accessed Jul. 25, 2022).

With a population of, 210,100 the London Borough of Camden is one of the 12 boroughs of inner London. The southern tip of Camden forms a small section of Central London.

The Council[5,6] will address social services policy planning in Camden by predicting risk factors across four distinct domains:

- housing and adult social care;
- children, schools, and families;
- culture and environment planning;
- building and parking services.

The Council is committed to the publication of open data and adheres to the Council Open Data Charter which is underpinned by the notion that residents have the right to access data which does not compromise individual privacy.

The plan is to apply predictive analytics techniques to data sourced from the target domains in order to identify risk factors across the full range of services offered by the Council. Examples of information to be obtained using these techniques include the following.

The extent of the relationship between housing repairs required and other higher cost services such as social work. Such information would enable the Council to identify potential cases where early intervention in one area could prevent cost escalation in another, related area.

- Car licensing data. Providing this information to the parking services department would enable planning for temporary events which might require additional resourcing.
- Data that enables forecasting of the numbers of people requiring adult social care services in future, principally older people, in order to plan future services.
- A comprehensive overview of the totality of each resident's interactions with the Council.
- The extent of the relationship between housing repairs required and other higher cost services such as social work. Such information would enable the Council to identify potential cases where early intervention in one area could prevent cost escalation in another, related area.

[5] "Open Data Policies for Citizens: Camden - Policy Cloud." https://policycloud.eu/pilots/open-data-policies-citizens (accessed Jul. 25, 2022).

[6] "Deliverables - Policy Cloud." https://policycloud.eu/publications/deliverables?field_wp_tid%5B%5D=7 (accessed Jul. 25, 2022).

Using the information harvested from data analytics, the Council will be able to design evidence-based policies and evaluate their economic feasibility, their political viability, and their legitimacy, thereby improving the quality of the services offered to citizens and achieving the efficiency and effectiveness which are key elements of good governance.

3.4 Trikala, Greece

Trikala was a demonstration site for the EU ELVITEN project – electrified L-category vehicles integrated into transport and electricity networks.

During the 6 months' demonstration period of the Horizon 2020 Project – ELVITEN, all trip data of the end-users (citizens and professionals) using EL-Vs (light electric vehicles) in Trikala were collected from the project's fleet management platform. The EL-Vs used were electric quad and three wheelers and electric bicycles and had special black boxes installed to gather data. All these anonymised trip data were transferred to the Municipality's open data platform in order to be available to companies, organisations, universities, and cities that are planning to implement mobility projects in the near future.

The types of data which were utilised included the following: trip data of electric vehicle users, vehicle speed, location data, trip times, battery level, and energy consumption.

The end-users of the EL-Vs were using a smartphone application in order to book the vehicle and to track and record all the critical data, such as vehicle real-time location, speed, time of trip, and energy consumption of every single trip they made. The application was also used to gather feedback through small questionnaires.

ELVITEN demonstrates how electric light vehicles (EL-Vs) can be used in urban areas and be integrated into the existing transport network of six European cities.

More detailed information can be found at their website.[7,8]

3.5 Umeå, Sweden

Through the RUGGEDISED project, Umeå[9] is implementing nine smart city solutions to ensure the city achieves two, often divergent, goals: population growth and a lessening of its environmental impact.

---

[7] "ELVITEN project." https://www.elviten-project.eu/ (accessed Jul. 25, 2022).

[8] "The ELVITEN EU project demonstrations." https://www.youtube.com/watch?v=JWl-5c6pyXFA (accessed Jul. 25, 2022).

[9] "RUGGEDISED - Smart city lighthouse project - UMEÅ." https://ruggedised.eu/cities/umeaa/ (accessed Jul. 25, 2022).

Often when trying to optimise energy use, the general approach has focused around supply chains and building logistics. Umeå has put a strong focus on its end-users – or citizens who have been ranked as some of the most active and environmentally aware citizens in Europe. But, there is still room for improvement! And the technology that is implemented is only as effective as its users. Therefore, a key aim is to encourage behavioural change, through gamification, as part of enabling a broader social shift in the adoption of more sustainable habits to all aspects of life.

Smart building management connected to the energy demand management system of different buildings in the University District, for which over 500 additional sensors have been installed, has been used to gain a better understanding of power consumption and to lessen the level of unnecessary energy use. On a district level, Umeå has developed different business models to enable buildings to share excess renewable energy and has implemented a system that makes it possible for buildings to store excess heat produced in the wider district. Work here has been useful in developing a more holistic view of the often complex business landscape of energy supply systems – local and distributed – and has provided the opportunity to pinpoint bottlenecks and find ways of mitigating them.

To make sure what follows the increase in population is not traffic jams and lower air quality, Umeå has implemented an innovative approach to parking for the more central parts of the city making it attractive for employers to support sustainable mobility. This, together with the installation of photovoltaics to power electric vehicles without adding burden to the overall energy system and a new climate-smart bus stop, are the approaches for better urban mobility tested in Umeå and have a high potential for replication in other cities.

While data is a universal language, translating it into local climate action remains a challenge. Umeå has developed an open-data decision platform (U8) allowing both city officials, outside experts, and citizens to access and visualise different data from the city; conduct market research in an out-of-the-box solution; and consider a process for presenting data consistently, whilst making sure the portal is sufficiently populated to be useful and ensuring developers of the future are aware of this data source. These are among a number of actions that Umeå has taken to ensuring that urban data is being used to achieve more sustainable, liveable cities.

In summary, the nine smart-city solutions are as follows:

- climate smart business model for 100% renewable energy supply;
- smart peak power control;

- gamification – influence behavioural patterns;
- intelligent building control and end-user involvement;
- climate smart bus station;
- smart city open-data decision platform;
- e-charging hub;
- smart business model for flexible parking;
- smart city open-data decision platform;
- demand side management.

3.6 Tampere, Finland

Fast growth of the city and its event sector and new emerging urban threats underline the need for innovative urban security solutions.

Utilising smart technology (sensors, data analytics, and situation awareness tools) enables:

- to integrate big event areas in the city centre with the surrounding city infrastructure;
- to support collaboration between event organisers, safety authorities, and third-party actors;

The main sources of data for this application come from traffic and analytical city cameras.

City cameras collect video footage from the city centre. The footage is streamed on a real-time basis to Insta Blue Aware (IBA) situation awareness platform[10]. IBA is a web-based tool providing a joint view to safety authorities, event organisers and event safety services, representatives of other safety operators, as well as to city safety management team. Based on the information provided by the camera stream, all actors safeguarding event safety and security can create operating models or act accordingly.

IBA product family includes the following: IBA situation awareness platform, IBT mobile app, and info link.

These solutions (solely or jointly) can be used:

- to support collaboration between authorities and non-authorities;

[10] "Insta - Front runner in secure digitalization." https://www.insta.fi/en/en/ (accessed Jul. 25, 2022).

- to enhance communication;
- to support building and sharing of the "big picture" and situation awareness;
- to enable picture/video sharing where no fixed cameras or other visual connections are available.

All these solutions are:

- easy and light, by using existing technology;
- offered by event city for smooth and safe events;
- scalable through increased amount of sensors, changing/evolving areas, and more functionalities.

AI algorithm that is developed in the Finnish SURE project[11] is able to detect disruptive behaviour from the video footage and in case of any, to send an alert to IBA platform.

3.7 Cities with Universities: KRAKEN and Students

Student data is protected by privacy-preserving cryptography and authenticated using self-sovereign identity solutions.

This pilot allows university students to trade their academic records in a privacy-preserving way, and recruitment agencies to acquire this data and process it, keeping the student's privacy intact.

Universities produce academic data of students, such as graduation certificates, certificates for each course, and the enrolment status for individual terms. With KRAKEN[12], a student can export their data into a cloud data wallet and offer it on a data marketplace. By using privacy-preserving cryptography, such as functional encryption and multi-party computing, a recruitment service provider can purchase a set of student data and use it for analytics, without invading the privacy of individual students. Additionally, KRAKEN uses self-sovereign identity solutions to always ensure user consent and a link between university/student and their data in a privacy-friendly way.

---

[11] "SURE! TAMPERE." http://suretampere.fi/ (accessed Jul. 25, 2022).

[12] "Education pilot - Kraken." https://www.krakenh2020.eu/pilots/education (accessed Jul. 25, 2022).

Objectives:

- to connect KRAKEN to real university student-information system (SIS);
- to define formats to export and share student data (grades and diploma) in a secure and privacy-preserving way;
- to allow authentication of students via eIDAS (or self-sovereign identities derived from eIDAS identities) to provide user consent;
- to allow authentication of student data by using a decentralised SSI ledger;
- to allow privacy-preserving data analytics on student data using FE and MPC cryptography;
- to demonstrate the applicability of KRAKEN architecture in general;
- to verify that design of KRAKEN components can be applied to a real-world use case.

Benefits and beneficiaries:

- Students: using student data for job applications and allowing to share and demonstrate data in a modular way without compromising the rest of the data.
- University: possibility to exchange student data while ensuring the students' privacy.
- Recruiter: making analysis of student datasets, e.g., gain insight into the academic performance of an individual student by comparing it with analytic of the dataset.
- Companies: facilitating the search for new talent during the selection processes.

Challenges:

Legal challenges: export of student data in a GDPR-compliant way, enabling user consent and privacy.

Technical challenges:

- connect the existing SIS to KRAKEN;
- define formats;
- combine the building blocks provided by KRAKEN to encrypt, authenticate, and analyse encrypted data;

- define data analytics system for the student data;
- connect to one or more data marketplaces;
- connect to the existing or new SSI ledger.

3.8 Rotterdam, Netherlands

One objective of RUGGEDISED is to transform large districts or buildings (old, new, and mixed) into low carbon and resource-efficient districts through smart interaction and integration of energy systems at district level. This requires smart management or infrastructures for electricity, heat, and cold as well as smart solutions for local energy production, storage, and exchange of energy. A representation of available open data of the district of Rotterdam[13] will be visualised on an open 3D platform, making it possible to monitor and communicate different information (starting with the energy performance of buildings), enabling endless applications and scalable to digital city level.

This solution will complement the "energy management system". Data on the energy use of the buildings, provided by the energy management system, will be matched and transferred into a new 3D city operations platform that will be developed.

This platform discloses and visualises the actual use of energy as well as use over a period of time (by individual buildings as well as the whole area). The 3D model is connected to the platform, and together with real-time data, it forms a 3D digital twin of the city.

This forms the basis for further innovation by making data available for everyone. The required infrastructure for this sub-solution is provided separately.

Both the city of Rotterdam as well as private parties will use the model for new applications, also to create new businesses. Connecting the many sensors and other data sources, such as municipal databases, requires standards for connecting to the data platform. The development of such standards is crucial.

Because the entire community will have access to data due to the use of open standards, it is expected that this 3D operations platform will become a powerful resource for further innovations.

The 3D city operations platform is an open data platform. So every third party can use data and visualisation tools to create their own business. This

[13] "RUGGEDISED - Smart city lighthouse project - ROTTERDAM." https://ruggedised.eu/cities/rotterdam/ (accessed Jul. 25, 2022).

will influence society and lead to additional economic benefits for companies and more wellbeing for citizens.

Once developed, the 3D city operations platform can technically be replicated in any city. However, there are a number of conditions for the development of the model and for maintaining effectiveness and availability:

- development and availability of open data standards;
- disconnecting data collection and data storage from application development;
- Data and platform ownership and governance issues;
- Avoiding vendor lock-in.

In addition, co-creation with commercial parties is required for acquiring knowledge and resources.

3.9 Athens, Greece

Athens is a global tourism destination as well as the largest metropolitan city and capital of Greece. Visitor numbers vary considerably with wide differences between the high and low seasons, and these variations have a knock-on effect on traffic within the city.

Athens is a demonstrator city in the DUET project[14] and part of the logic for joining the project was that whilst the city has an overall plan to transform the city, concentrating on digitalisation of services to improve them for their citizens, the city does not have access to vast amounts of readily usable and well-managed data covering the city. So the potential for using a digital twin of the city was recognised, which would have the capacity to bring all of the city's digital sources together and become more accessible. This would create the opportunities for experimentation and exploration in order to maximise the potential in running the city. Athens sees data-driven decision-making as a key pillar for city and business transformation particularly in light of the rapid changing and challenges being faced. Athens is in the early stages of its digital transformation roadmap and is starting from scratch.

Thus, its practical goals in adopting the "digital twin" approach through the DUET project can be expressed as the city wishing to:

- understand city relationships and overcome engagement barriers with stakeholders;

[14] "Athens Twin - DUET: Digital Urban European Twins." https://www.digitalurbantwins.com/athens-twin (accessed Jul. 25, 2022).

- create new business value based on data-driven insights;
- co-create digital services with the active engagement and participation of citizens;
- generate decision-making approach using common standards for greater interoperability of digital tools;
- improve effectiveness of policy design and implementation.

In relation to the city's policy goals, a driving force behind using digital twins is to help achieve the city's overall goals and, in particular, the issue of reducing air pollution.

"Athens is cracking down on air quality, around six per cent of all deaths in Greece are linked to air pollution, towards which diesel engines are acknowledged as a leading contributor, along with a high smoking percentage of the population. With nearly 30 percent of Greece's population living in the capital, having government departments working on transport, health and environmental policies to tackle the issue together seems eminently sensible".

3.10 City Health Organisations and the KRAKEN Health Application

All cities need to care for the health of their citizens and the KRAKEN health and wellness data marketplace for individuals and organisations can assist.

Leveraging the established data platforms of the H2020 MyHealthMyData (MHMD) and Streamr projects, KRAKEN[15] will develop a biomedical data marketplace for individual citizens and healthcare organisations, where it will be possible to commercialise patient medical records and real-world, wellbeing data streams from wearable devices, in full compliance with GDPR.

KRAKEN is developing a biomedical and wellness data marketplace that connects individual citizens and healthcare organisations and shares medical and wellness data streams with data consumers, such as academic research centres, health-tech companies, insurers, public authorities, and wellbeing service providers, in exchange for economic value.

The trading will involve personal health records including lab results, medical histories, and radiology images, and health and wellbeing, real-world

[15] "Health pilot - Kraken." https://krakenh2020.eu/pilots/health (accessed Jul. 25, 2022).

data such as heart rate, dietary, and physical activity, recorded by mobile apps and other wearable devices.

Objectives:

Developing a technical solution enabling citizens and healthcare organisations to:

- fully control what data is shared, with whom, and for what purpose;
- make the data from their devices discoverable to interested parties;
- deliver their data from their devices to those who want it in a simple way;
- receive a share of the value of the data they are producing.

Benefits and beneficiaries:

- Data providers: individual citizens and healthcare centres (hospitals) will be able to control and regulate the use of the data they generate and be equally rewarded for that, increasing awareness of their rights and liabilities as data controllers and sharing the economic benefit deriving from data usage.
- Data users: academic research centres, health-tech companies, insurers, public authorities, and wellbeing service providers. All of them will be able to gain access to high-quality aggregated data for their work in an unprecedented straightforward and seamless way, being able to extract maximum value from data, in turn benefiting society by creating added economic value as well as accelerating the pace of data-driven research and technological innovation in the biomedical field.

Challenges:

- achieving seamless aggregation, curation, and harmonisation of heterogeneous data sources;
- reaching full data interoperability between different systems;
- maximising data discoverability by implementing an efficient data catalogue.
- developing user experiences and interfaces intuitive enough to allow effective utilisation by non-expert users.

3.11 Sofia, Bulgaria

Urban policy-making through analysis of crowdsourced data:
Sofia is the capital of Bulgaria and the biggest political, administrative, cultural, and educational centre in the country, with a current population of 1.8 million inhabitants. The municipality of Sofia is constantly working to improve the urban environment and meet the challenges that the city is facing. Evidence-based policy-making is crucial for addressing urban challenges in a cost-efficient way.

Through its involvement in Policy Cloud[16], the Sofia municipality will address urban policy as a critical success factor in improving the overall urban environment of the city. Policy design will be adapted based on analysis of big data sourced from the following sectors:

- transport, parking, and road infrastructure;
- waste collection and waste disposal;
- cleanliness of public spaces;
- ecology and green systems;
- violation of public order;
- others, of importance to citizens.

In designing policy, we will combine the data from both existing sources and from new open datasets that become available.

The existing data source is the citizens' contact centre, which has been operational since 2014 and facilitates direct communication from citizens, industry, and institutions wanting to signal non-urgent deviations from normal practice within the urban environment.

Analysing the territorial distribution of these signals by category will enable municipal and district administrations to identify problems, issues, and behaviour trends in the urban environment. The analysis will also facilitate monitoring and control of the services under review, enabling preventative action to be taken where potential risk is identified, and guiding decision-making

[16] "Urban policy making through analysis of crowdsourced data - Policy Cloud." https://policycloud.eu/pilots/urban-policy-making-through-analysis-crowdsourced-data (accessed Jul. 25, 2022).

around policy adjustment and/or adoption and also around the effective use of budget and public resources.

Crucially, where policy is introduced or changed as a result of the insights obtained, data analytics will enable the authorities to understand the effects of the change and find explanations for the behaviours observed.

Policy Cloud big data streaming and real-time big data platform will enable the Sofia municipality to improve operational efficiency, transparency, and decision-making.

3.12 Piraeus, Greece

Large commercial districts are vulnerable for both inbound and outbound smuggling, street and organised crime. The city of Piraeus, through the Urban Innovative Actions (UIA) project BeSecure-FeelSecure (BSFS)[17,18], aims to reinforce urban security and promote positive perception of urban safety by providing strategies and tools to link the main urban security stakeholders and facilitate their collaboration in physical-and-cyber space. BSFS offers, among others, a digital platform called Collaborative Urban Risk Management platform (CURiM), enabling the assessment of combined physical cyber threats and the visualisation of risks and other relevant activities through geospatial mapping, to facilitate decision-making. To do so, a wide variety of information is gathered and exploited.

The types of data utilised include:

- information and statistics of crime from the Hellenic Police reports;
- feedback from people through the CURiM mobile application;
- web mining and open intelligence sources;
- sensors.

The evidence-based Collaborative Urban Risk Management (CURiM) ICT platform gathers all available information from multiple sources (CURiM app, OSINT, sensors, and police reports) and facilitates crime risk assessments on selected areas of the city of Piraeus. The analysis investigates the criminality status of the area and helps the decision-making process of the

---

[17] "Be Secure Feel Secure (BSFS) - Piraeus." https://www.bsfs-piraeus.eu/ (accessed Jul. 25, 2022).

[18] "BSFS Video (English version)." https://www.youtube.com/watch?v=gOP-9bO5LKE&t=4s (accessed Jul. 25, 2022).

responsive action plans that should be followed, in order to mitigate the criminality in the city of Piraeus.

A visualisation component presents/depicts risk assessment metrics providing useful analytics for its users. The CURiM system helps estimate the probability of different urban crimes in different areas of the city.

A crime mapping service/component can visualise the information gathered into a crime heat map for any selected area within the city of Piraeus, thus providing a basic view of the current status as well as the evolution of crime metrics in time.

The information extracted from CURiM is then provided to the Local Council for Crime Prevention (LCCP), recently established in the city of Piraeus. The LCCP brings together local authorities in collaboration with other urban stakeholders to discuss these reports and use this information to prioritise urban security requirements, strategy, and allocation of resources.

3.13 Grand Lyon (Metropolis of Lyon), France

The Metropolis of Lyon is experimenting the My Data[19] approach to allow citizens to monitor their own energy consumption for saving money and reducing their personal environmental impacts. Data is provided by different providers and is stored in the citizen's personal cloud[20] where no external subject can access.

Citizens can store their data related to energy consumption, coming from different operators (gas, water, electricity, etc.) on a personal cloud that can be accessed only by them. From such data, citizens are able to monitor their own consumptions and modify their own attitude to save money and reduce environmental impacts. No external subjects can access citizen's data and the administration is not able to manage it directly: the objective is to make citizens responsible towards the climate change issue by providing them all the necessary information to manage their personal environmental impact.

The access to the personal cloud is free for citizens and is paid by the administration. The service is provided by Cozy[21] which gives to every interested citizen a unique entry point to all available data.

[19] "MyData." https://www.mydata.org/ (accessed Jul. 22, 2022).

[20] "Ecolyo." https://ecolyo.com/ (accessed Jul. 25, 2022).

[21] "Cozy Cloud - Prenez le contrôle de toutes vos données." https://cozy.io/fr/ (accessed Jul. 25, 2022).

The same approach can be extended also to other services, such as finance, insurance, and health.

3.14 Prato, Italy

The experience was carried out in the context of the EU H2020 Project "Raising Open and User-friendly Transparency-Enabling Technologies for Public Administrations (Route-to-PA)".[22] The project objective was the setup of ICT-based tools to improve the engagement of citizens in the open data environment by forming or joining existing online communities that share a common interest and discuss common issues of relevance to local policy, service delivery, and regulation. The project tools allowed citizens both to access open data to activate discussion on public policies based on such data and to produce open data themselves to enrich the city information framework.

The city of Prato tested the project approach in the enlargement of the city Wi-Fi network, by asking citizens to build an open dataset to locate new Wi-Fi access points and to support their proposal through a data-based discussion on the project platform. The result was adopted by the administration as a basis for its policy of improvement of the city Wi-Fi network.

Types of data utilised:

While planning the pilot activities, it was assumed that discussions on this topic might be influenced by several factors, like the population distribution by age and nationality and the dislocation of aggregation points such as sports facilities, libraries, schools, parks, and green areas.

Starting from this assumption, the administration made available several open datasets on the city CKAN platform[23] to support citizens' choices:

- location of the current Wi-Fi network antennas;
- population distribution by age and elementary statistics units (the city is divided into 34 ESU representing the city zones);
- foreign population distribution by range of age and elementary statistics units;
- location of schools;
- location of green areas;

[22] "ROUTE-TO-PA - A Horizon 2020 project." https://project.routetopa.eu/ (accessed Jul. 25, 2022).

[23] https://ckan.org/

- location of libraries;
- location of squares;
- post offices (they are equipped with their own Wi-Fi spots);
- national broadband coverage maps.

Description of the application:

The users logged in the project platform and there could access a specific room where the provided datasets were available for analysis and visualisation, but they also had the possibility of searching autonomously for other information to support the discussion. Through the co-creation feature, the users were also able to build one or more new datasets containing the position and description of the proposed new Wi-Fi locations and to eventually produce a shared document to present their proposal to the administration as a contribution to the Wi-Fi network enlargement.

A moderating role was assumed by the administration, particularly to give information and clarifications if necessary.

The final result was an open dataset collecting all the citizens' proposals that was also added to the open data platform and used by the administration to widen the city Wi-Fi network.

3.15 Eilat, Israel

Eilat is one of the associated counties complementing the 100 net-zero cities.

Eilat's smart city team has developed in collaboration with the chief scientist of the Ministry of Energy, and a local start-up, "Solview", a unique one-stop-shop platform for solar PV installation. It gives each house owner a rooftop view of a financial prediction of solar PV installation.

The data includes calculating the potential surface, expenses, return on investment of future electricity production, finding suppliers, and funding opportunities.

The pilot has ended, and the project is now in the process of updating national and international permits and licenses.

With the help of the citizens' own properties, data, and active participation, this innovative platform will substantially increase the number of efficient PV systems in the city, thus democratising energy and reducing costs and fossil fuel usage.

More detailed information can be found at their website[24].

---

[24] "Eilat Smart City." https://www.eilat.muni.il/%d7%90%d7%99%d7%9b%d7%95%d7%aa-%d7%a1%d7%91%d7%99%d7%91%d7%94/ (accessed Jul. 25, 2022).

3.16 Florence, Italy

The city of Florence[25,26] has been very active for years in using data to drive decisions and take care of the citizen wellness and environment. The global vision is to employ all currently available data coming from sensors, internal database, and users in order to describe and monitor the city status in every moment; one place to collect, process, and deliver data and information to all the different decision-makers.

The city refers this as virtual smart city control room[27] and a consistent part of it is dedicated to the environment protection.

Even if our data are generated by people's behaviour, they cannot be intended as "personal" data. Some data are coming directly from IoT sensors and others are voluntarily provided by the user through using our city apps[28]. Anyhow, all data that the city uses are every time anonymous and processed in aggregate form.

Data sources include:

- IoT sensors;
- city smartphone application data;
- data exposed from other city players/utilities through API;
- internal data (DB).

The application described relates to tourism: the city uses data of public Wi-Fi network, metro, and city events to monitor and predict the crowding level and prevent congestions in many relevant places inside the city centre. Other than for people, the promotion of the so-called "diffuse tourism" is good for the environment too. The city can better plan and increase efficiency in resource consumption (e.g., improving the availability of public transport just when and where it is actually needed).

Mobility data analysis: coming from different mobility sensors (e.g., number of vehicles in and out from the city, free parking spots available,

[25] "Firenze digitale." https://www.firenzedigitale.it/ (accessed Jul. 25, 2022).

[26] "#IF2021: la Toscana presenta Smart Region, l'arma in più per gestire i big data - intoscana." https://www.intoscana.it/it/articolo/if2021-la-toscana-presenta-smart-region-larma-in-piu-per-gestire-i-big-data/ (accessed Jul. 25, 2022).

[27] "Snap4City per la creazione di Smart City Control Room, SCCR (ITA)." https://www.snap4city.org/drupal/node/427 (accessed Jul. 25, 2022).

[28] "Smart city & IoT: più dati e meno emissioni." https://www.youtube.com/watch?v=My2dblBnEVM&t=109s (accessed Jul. 25, 2022).

etc.) in order to estimate presences and road congestion levels. With this non-personal information, decision-makers are able to take actions (over a short and medium term) that can improve the efficiency of the entire mobility system (public and private) inside the city.

Sharing vehicles position analysis: the position of the new sharing electric mobility vehicles (electric bike, scooters, and cars) in order to monitor the distribution across the city and consequently better allocate them to improve their usage.

Through the Infomobilità Firenze (IF) app users can also contribute to share their travel habits in terms of departure and arrival places, preferred vehicle, and street paths. App users can also see the real-time map of all available sharing vehicles and choose to rent one. Once collected, these data help the city to meet the citizen mobility demand, increasing the efficiency and reducing total emission amount.

The project Firenze Green Smart City: an innovative digital project composed of multiple initiatives like:

- a comprehensive digitalised and interactive map of the city green areas;
- smart irrigation: a group of city gardens where the irrigation is completely and automatically handled by IoT sensors;
- a digital platform where citizens can choose to donate new plants to the collection.

Together, these digital platforms, as well as improving or offering new services for citizens, generate new data contributing to better manage, develop, and protect urban green areas.

In relation to the tourism and mobility field, the city is also experimenting on its big data platform with some machine learning models, in order to continuously improve the prediction phase by taking advantage of all the data sources that the city has.

3.17 SmartEnCity: Vitoria-Gasteiz, Spain

One of the three lighthouse demonstrators implemented within the SmartEnCity project is the city of Vitoria-Gasteiz, Spain[29]. An urban management system was designed as the platform where all the information

[29] "Vitoria-Gasteiz urban management system - SmartEnCity.eu." https://smartencity.eu/about/solutions/vitoria-gasteiz-urban-management-system/ (accessed Jul. 25, 2022).

generated in the city can be stored, processed, and retrieved for providing management solutions in the areas of energy efficiency for buildings, urban mobility, and citizen engagement.

The core of the platform includes components like data acquisition layers, data models, data real-time repositories, security, etc. Those data are obtained using low-level sensors measuring temperature, relative humidity, CO_2, gas and energy consumptions, as well as on-board acquisition devices installed on the electric buses.

On the other hand, a city platform was implemented to store and analyse the data gathered from the city and including services offering information on energy consumption and comfort conditions. Citizens have at their disposal specific applications to interact with the platform in a transparent way providing information based on real-time data as comfort conditions, impact calculation of the energy retrofitting of the houses, local news of interest, forecast of energy demand, and high-level visualisation of indicators and key performance indicators online.

3.18 SmartEnCity: Tartu, Estonia

Another example of the implementation of the SmartEnCity concept was developed in Tartu city[30], consisting of the installation of smart home systems in each of the apartments of a specific area, for collecting energy consumption data. The data will be used for improving the configuration of heating and ventilation devices in the renovated buildings, providing end-users with direct feedback about their consumption habits and for building up innovative services by third parties, always assuring the privacy of the end-users.

The technical components of those smart home systems include gateways for communication with the meters, sensors and devices, tablet control panels, smoke detectors, impulse counters, smart thermostats, and CO_2 detectors.

The benefits that the citizens obtain, among others, are the possibility of monitoring and adjusting energy consumption, improved data availability and new business opportunities, increased comfort and control over the indoor climate, behavioural change related to consuming energy, increased energy efficiency and reduced energy bills, greater transparency of urban processes, and better management of service providers.

[30] "Smart home solution (Tartu) - SmartEnCity.eu." https://smartencity.eu/about/solutions/smart-home-solution-tartu/ (accessed Jul. 25, 2022).

3.19 Helsinki, Finland

Digital twin and air quality:

The digital twin for mobility is part of the development of the city of Helsinki's digital twin and the digitalisation of the city in general[31]. The urban environment is a complex entity, with data on it scattered across a large number of different information systems. Therefore, the digital twin is a "system of systems" that can contain many different views that serve different purposes. Different use cases require different input data and different tools to make use of that data. New and even surprising combinations of data offer the opportunity to provide added value and new understanding. The full potential of the urban digital twin is not yet known.

The key benefit of the digital twin is the approach it offers: before major changes are made to the urban environment in the real world, the effects of these changes can first be tested digitally. This saves resources and helps build better lives in cities, as the urban environment can be improved through informed decision-making.

The UIA HOPE project explored how urban residents can participate in air quality measurement and how companies can make innovative use of air quality data. Residents measured local air quality in Vallila, Jätkäsaari, and Pakila.

The road dust season has started and air quality is poor in places. Even in Helsinki, limit values are exceeded every year. Traffic is the largest source of emissions.

In Helsinki, 157 volunteers measured air quality in their residential environment using portable sensors in Jätkäsaari, Pakila, and Vallila. The automated sensors used by the residents carried out more than a million air quality measurements during the research period. The measurements were used to determine whether the residents could be involved in air quality measurements and how reliable the results can be obtained by means of crowdsourcing. The measurements also increased the understanding of air quality among the residents. Through the participatory budgeting method, residents were able to vote on measures to improve air quality at the urban level. Furthermore, on the basis of the measurements, residents were able to change their daily routines so that they used routes with cleaner air.

[31] https://www.hel.fi/helsinki/en/administration/information/general/3d/3d

3.20 Glasgow, Scotland

(Associated Country "climate-neutral city")
Smart open data decision platform and central management system:

This solution includes the creation of a query based geo-spatial "data-based decision platform" (DBDP)[32] that will collect data related to city management (e.g., energy, air quality, traffic flow, etc.) and provide analysis of multiple datasets to enhance strategic planning in the city (including energy planning). Glasgow City Council will utilise the existing open data platform and build DBDP around existing ICT infrastructure. This represents a technical challenge.

The platform will sit on the council's existing platform, gathering city management data. This data will be combined with existing datasets to provide insight into the impact of these smart city operations on services delivered by the council to enhance people's lives. Furthermore, through the monitoring of impacts, targeted interventions can be made in priority areas across the city.

The solution should:

- provide a means to understand the impact of smart city interventions through real-time visualisations;
- provide a means to combine data from smart city interventions with static datasets within the council;
- help non-data "experts" explore the data – real time or otherwise – to help with city planning, stakeholder engagement, etc.

The creation of a DBDP will be an open system for other developments in the city. The platform upon which the tool is built is already at city scale so that little further investment will be required.

[32] "RUGGEDISED - Smart city lighthouse project - GLASGOW." https://ruggedised.eu/cities/glasgow/ (accessed Jul. 25, 2022).

4

Case Studies Involving the Use of Personal Data in a Smart City

Paolo Boscolo[1], Michail Bourmpos[2], Ilesh Datani[3], Ana García Garre[4], Vincent Keunen[5], Borja Molina-Ríos[6], Elena Palmisano[7] and Christina Tsiligkiri[8]

[1]Comune di Prato, Italy
[2]IT Department, City of Piraeus
[3]Assentian, Ireland
[4]MIWenergia, Spain
[5]Andaman7, Belgium
[6]MIWenergia, Spain
[7]Comune di Prato, Italy
[8]Olympiacos SFP, Piraeus, Greece
Email: [1]p.boscolo@comune.prato.it; [2]mmpour@piraeus.gov.gr; [3]ilesh.dattani@assentian.com; [4]ana.garcia@miwenergia.com; [5]vincent.keunen@andaman7.com; [6]borja.molina@miwenergia.com; [7]elena.palmisano@gmail.com; [8]ctsiligkiri.erasmus@olympiacos.org

Abstract

This chapter, first, provides examples of how the evolving technologies, which enable the safe and secure sharing of personal data, under their own control and for re-numeration or altruistic purposes and with whom they choose, have been demonstrated. They range from usage in smart city environments through to that in an energy company and in the health arena.

And, second, it attempts to illustrate how these technologies and applications can be used to augment and give added value to those examples of the general usage of data within a smart city. We have tried to show that by safely adding the personal data coming from citizens, we can also improve the original scenarios described and augment them in a positive way.

DOI: 10.1201/9781003399384-5

It is not in any way attempting to be a definitive guide but merely to show what is possible and worth further investigation.

4.1 MIWenergia in the DataVaults Project

MIWenergia is an SME electricity retailer founded in Murcia, Spain. It operates in Spain nationwide with a varied range of customers: residential, industries, small and medium enterprises, and large tertiary buildings.

Providing service to more than 3000 customers with an annual energy transit of 100 GWh, MIWenergia offers clients additional services related to energy efficiency such as energy consulting and audits, design and installation of photovoltaic systems for self-consumption, electric vehicle charging stations, and energy monitoring.

The first step of all methodologies used to provide the previously mentioned services is collecting the needed data to be processed. Some data are public and others are technical ones and are indeed in the hand of the service provider, but the most important data are only known and owned by the user, such as the history of electricity consumption, number of inhabitants in the house, or the building orientation.

There are currently payment platforms for third-party access to electricity consumption data. These platforms provide monthly aggregated data per supply point; so the information is not very detailed, being almost useless for the design of a photovoltaic system or the energy prediction. A platform like DataVaults, where all the user-generated data is provided to the platform by the data owner, can give access to all the companies interested in those data, connecting the data processors, distribution system operators, or electricity retailers, and letting other energy entities use that detailed information about electricity consumption with the user's permission. MIWenergia considers a platform for accessing data like DataVaults needed due to the low quality of the platforms and data that are available now. Furthermore, the company only has access to the data of their customers; DataVaults gives the opportunity of having the data of all the platform users. MIWenergia will utilise DataVaults, to try and acquire such data directly from customers that are willing to share those, towards providing better services to them.

The role of MIWenergia in the DataVaults project is double: first, MIWenergia acts as a data provider, helping to connect the platform with their database of hourly energy consumption. Second, MIWenergia act as data seeker, testing the usability of the platform related to developing new energy services or designing better-targeted marketing campaigns.

As data providers, the data owners registered in the DataVaults personal app can share their energy consumption data with the data seekers through the MIWenergia connector. The connector accesses the MIWenergia server and collects the data. The data is collected once a week and the time interval of the data is 1 hour. The MIWenergia connector can obtain the data of our clients automatically. The client only needs to introduce her MIWenergia credentials when connecting the energy consumption source. However, if they are not our direct clients, they will need to give us permission to access their energy consumption data. With their consent, we can provide the energy consumption data of all Spanish supply points.

As data seekers, MIWenergia (or any energy services provider) can be interested in obtaining the energy consumption profiles of non-clients through the DataVaults platform in order to offer new and innovative services and targeted marketing campaigns. Within the DataVaults project, we are testing three different cases of study towards this objective.

The main goal of the first scenario is to improve the design process of a PV system for self-consumption. Normally, when designing a PV system, the companies do not have access to the energy consumption profile of the clients. In this way, the companies design the PV system based only on total consumption data, provided by the client, and then estimate the hourly consumption profile and the potential for self-consumption. Having the real historical hourly energy demand "automatically", it is possible to make more precise calculations about the percentage of self-consumption and the optimum size of the PV system. For the optimisation of this design, the DataVaults platform can be very useful, as the clients can easily share their energy consumption profile. In addition, on the same platform and at the same time, they can provide other useful information such as the location and orientation of their dwelling, available space, type of roof, etc. In this way, the process of designing the PV system is simplified and efficient.

The second case of study is related to making energy consumption classification for dwellings and sending tips to the platform users to improve their energy efficiency. When a data owner shares with the data seeker their consumption data, MIWenergia can use this information to calculate some energy-related indicators and classify their dwelling in terms of energy efficiency. In this way, MIWenergia can send them suggestions about measures and energy efficiency tips that they can follow to reduce their energy consumption and their electricity bills. At the same time, this information can be useful for the company to have the profile of the "worst" dwellings in terms of energy efficiency to design specialised energy efficiency services specifically to these sectors or profiles.

Finally, the last scenario to test is related to the demand prediction and the interests, hobbies, and habits of the customers. Having the energy consumption profile, we can predict some habits of the consumers and design targeted marketing campaigns for energy services specifically to them. For example, if we found that customers under 30 years consume more energy on the weekends, then the company would be able to design a cheaper electricity tariff in this period and a marketing campaign oriented to this profile to engage them. In addition, other companies not energy-related can also design marketing campaigns in this way. For example, if we detected families with children reduce their energy consumption on Friday nights, then a restaurant would be interested in designing a special offer for this day to these kinds of families (assuming they leave the house for leisure). Energy consumption profiling can be very useful for many marketing applications.

These scenarios are closely related to increasing the consumption from renewable sources and improving energy efficiency; so the private sector can contribute to the European goal of net-zero emissions cities. However, we need to consider that energy consumption data has much more possibilities to be used to achieve this goal and public institutions and councils can take advantage of all this information for designing smart and carbon net-zero cities. In the following sections and based on existing projects, we are going to propose some other applications in which DataVaults can be useful for the cities to access energy consumption data and add value to smart and carbon net-zero cities.

This experience can provide added value to many of the examples of current usage of data set out above in Chapter 3. For example, that of *Eilat, Israel* (Section 3.15): designing PV installations. Eilat's smart city team has developed in collaboration with the chief scientist of the Ministry of Energy, and a local startup, "Solview", a unique one-stop-shop platform for solar PV installation. It gives each house owner a rooftop view of a financial prediction of solar PV installation. The data includes calculating the potential surface, expenses, ROI of future electricity production, finding suppliers, and funding opportunities.

The DataVaults platform can collaborate with "Solview" to provide energy consumption data of the citizens which helps to optimise the designing process of a PV solar system in a similar way to the MIWenergia first case of study.

Another example of how improvements can occur by providing access to a citizen's personal data can be seen with Glasgow's "*Creating energy maps*" (Section 3.20). The city of Glasgow has created the "smart open data decision platform & central management system". This solution includes the creation of a query-based geo-spatial "data-based decision platform" (DBDP) that will collect data related to city management (e.g., energy, air quality, traffic flow, etc.) and provide analysis of multiple datasets to enhance strategic planning in the city (including energy planning). The Glasgow City Council will utilise the existing open data platform and build DBDP around the existing ICT infrastructure.

The DataVaults platform can contribute to the DBDP by providing any personal data from citizens willing to share it. In the case of energy planning, users of the DataVaults platform can provide the energy consumption of their dwellings. This information can be useful for creating an "energy map" of the city, based on the demand of the citizens. In this way, knowing where and how much energy is being consumed in a city, the council can design specific actions for these places in which there is more energy consumption.

For example, if in a specific neighbourhood there is high consumption at midday, the council can plan to install a PV system nearby in a public space and create an energy community. People (and companies) can be interested in being part of the energy community and reducing their energy consumption in an easy way and with less investment than having their own PV system (only paying a fee for being part of the community). DataVaults can act as the platform in which citizens share their energy consumption to create the energy map and, at the same time, can be used by the council to ask for the interest of being part of energy communities when designing it. Besides, the process to apply for being part of the energy communities can be integrated into the DataVaults platform.

Similarly, advantages can be foreseen for the *Trikala, "Creating a "storage map" of a city*" (Section 3 4) application. Trikala was a demonstration site for the EU ELVITEN project – electrified L-category vehicles integrated into transport and electricity networks. All trip data of the end-users (citizens and professionals) using EL-Vs (light electric vehicles) in Trikala were collected from the project's Fleet Management Platform. All these anonymised trip data were transferred to the municipality's open data platform in order to be available to companies, organisations, universities, and cities that are planning to implement mobility projects in the near future. The types of data that were utilised included the following: trip data of electric vehicle users, vehicle speed, location data, trip times, battery level, and energy consumption.

The ELVITEN project is focused on the mobility part of the EV, but electric vehicles are also batteries that can be seen as a potential resource of the electricity grids. In this way, councils can be interested in creating a real-time "storage map" of the city. In conjunction with the information already registered by the ELTVITEN project, information on batteries already available (not only batteries from EV but also other "steady" batteries from dwellings or industries) and electricity consumption from the citizens can be shared through the DataVaults platform. All this data can be used to improve energy efficiency and manage the variability of renewable resources. A practical example can be a citizen who has an electric car with a battery of 60 kWh. The citizen has a work schedule from 8 am to 3 pm and, during this time, his car is parked nearby his workplace. He needs the car to go back to his home at 3 pm, which is 20 km far; so he needs enough battery at 3 pm for this trip. During the rest of the hours, the battery can be connected to the grid and be manageable as the council (or distribution system operators (DSO)/ transmission system operators (TSO) wishes to increase the energy efficiency of the grid. DataVaults can be the platform in which people can register for these demand response actions (so they can say that they are willing to participate) and share the necessary information, or give the necessary permissions, to access their data in exchange for some incentives from the council or DSO/TSO (that act as data seeker). All the batteries can be aggregated and managed in a more efficient way than with the conventional individual use of the citizens.

Improving the performance of the electricity grid and the integration of renewable resources with demand response actions through batteries in Trikala will contribute to achieving their goal of net-zero emissions.

Further, a contribution could be foreseen to the *Rotterdam, "Integrating energy and storage maps"* (Section 3.8) initiative.

One objective of RUGGEDISED is to transform large districts or buildings into low carbon and resource-efficient districts through smart interaction and integration of energy systems at the district level. This requires smart management or infrastructures for electricity, heat, and cold as well as smart solutions for local energy production, storage, and exchange of energy. A representation of available open data of the district will be visualised on an open 3D platform, making it possible to monitor and communicate different information (starting with the energy performance of buildings) and enabling endless applications and scalable to the digital city level. This solution will complement the R8 "energy management system". Data on the energy use of the buildings, provided by the energy management system, will be matched and transferred into a new 3D city operations platform that will be developed.

The DataVaults platform can complement the open 3D platform for Rotterdam, helping to create the energy and storage maps that have already been described for the previous cities (Glasgow and Trikala). Citizens from Rotterdam registered in DataVaults can share their consumption data and the information about their PV systems, batteries, and/or electric vehicles in order to integrate energy and storage maps in the RUGGEDISED 3D platform. They can also apply for being included in demand response programmes. In this way, all the energy services available in the city can be managed in an aggregated way, increasing their performance and reducing the emissions of the city.

And similar benefits might be created with additions to Umeå's "Smart Demand Management" (Section 3.5). The DataVaults platform can complement the work done in Umeå within the RUGGEDISED project, helping to create the energy and storage maps that have already been described for the previous cities (Eilat, Glasgow, and Trikala). Citizens registered in DataVaults can share their consumption data and the information about their PV systems, batteries, and/or electric vehicles in order to integrate energy and storage maps in the RUGGEDISED open platform. They can also apply through DataVaults for being included in demand response or demand side management programs or other new business models based on sharing and or storing the surplus of renewable energy. In this way, all the energy services available in the city can be managed in an aggregated way, increasing their performance and reducing the emissions of the city. The information about electric vehicles can also be useful for planning new charging stations to support sustainable mobility.

A further example of how utilising a citizen's personal data can contribute to *carbon-neutral cities* is in the example where a city is encouraging citizens to change their behaviour. In many cities, it is very common for municipalities and councils to offer grants for energy efficiency measures, such as changing lighting fixtures or photovoltaic systems. Normally, it is tedious to apply for them; so people end up not asking for them or having many problems in the process. In addition, justifying that efficiency measures have had a real impact could be difficult as the municipality does not have easy access to real energy consumption data of the citizens. The DataVaults platform can be required by the administration to make a simplified application. In this way, public entities would have access to real data verified by a platform, and the client would have easier access to subsidy management. In addition, in some places, it is compulsory to have an energy efficiency certificate of the dwelling to ask for grants or for selling a house. Councils can manage all the processes of this certificate (or other energy-related certificates) through the DataVaults platform, reducing the time and efforts spent by the council and by the citizen, as the energy consumption data and the rest of the information needed can be shared at once through the platform.

Energy consumption data in dwellings can be used to know if everything is okay in a house. By predicting energy consumption and monitoring it in real time, we can compare and analyse the deviations. These deviations can be signals of something "not normal" happening. For example, in the PHARAON (H2020 project),[1] MIWenergia is monitoring with IoT devices the energy consumption (and other parameters) of older people who live alone to detect if something wrong is happening to react faster to potential health problems. For example, if the citizen's energy consumption is always increasing from 11 am to 1 pm because the person is preparing lunch and one day there is not any increase, an alert is shown, and they need to push a button in the app to check-in that they are okay. The DataVaults platform can be used in order to reduce the number of monitoring devices needed, as you can easily share the health and energy information with the public services in the same environment. The citizens can apply through the platform for being included in the monitoring programs from the health institutions. At the same time, the health institutions can use DataVaults to reach the potential patients to be included in the program.

4.2 Prato's Usage of a Citizen's Personal Data

Smart cities are now adopting many technological solutions able to ingest and manage a large variety of data, ranging from geographic to urban, statistical, climate, and so on, according to the big data paradigm which is able to provide different and complementary data insights to support effective planning solutions.

On the other hand, this framework is somehow missing the adoption of citizens as data source, able to provide their personal data as an additional asset to improve the public administration's knowledge on the city context. Whenever personal data is required in the planning activity, the city administration needs to turn to third parties, such as ICT operators or social networks, to pay large sums to obtain what its own citizens are generating, even without them being aware of it and getting no advantage for their contribution. The city of Prato is very interested in exploiting the opportunity of directly interacting with citizens to get their personal data in a safe, secure, and privacy-respectful way and even to provide them a reward: this will help the administration to strengthen the relationship with citizens and to reduce costs for data acquisition. The participation in the DataVaults project has been a strong incentive to such approach and three different scenarios have been conceived to foster the integration of citizens' personal data in the information framework managed by the city.

[1] "PHArA-ON - Pilots for Healthy and Active Ageing in Europe." https://www.pharaon.eu/ (accessed Jul. 25, 2022).

The first scenario is related with mobility: through the DataVaults tools, citizens will be able both to share their location and mobility preferences to augment the provision of answers to specific questionnaires sent by the administration, by getting a compensation for their data. This will enrich the information framework and improve the planning capability of the Mobility Office in the city. The possibility offered by DataVaults to get personal data (including answers to questionnaires and surveys) tuned on mobility behaviours will improve the analysis of local mobility also on the basis of short-term evolutions. In addition, this will also improve the quality of surveys and questionnaires, due to the possibility of building citizens' samples that are more coherent with planning objectives. Eventually, the DataVaults approach allows to lighten the administrative burden in managing the users' privacy since the data remains always in the full possession of individuals who are free to decide if and how to share it. On the other hand, the implementation of compensation schemas will facilitate the building of a trust and valuable relationship with data owners that will encourage them to share their personal data and participate in questionnaires and surveys pushed through the DataVaults platform.

The second scenario deals with culture: the city museums will benefit from citizens sharing data on their cultural interests, location, and participation in cultural events, to improve marketing activity and better target it on specific user categories. The DataVaults platform can be used by cultural institutions for marketing surveys and loyalty campaigns for users of cultural services, identifying samples of individuals to whom they can send information material and ask for feedbacks, also including the provision of benefits (remuneration), such as example discounts, free tickets, etc. In addition, cultural institutions can use the DataVaults platform to collect information on people's cultural consumption and build more effective questionnaires and surveys to be pushed through the platform. Individuals will be able to share answers and feedbacks according to the data sharing and compensation procedures implemented by DataVaults and this will hopefully increase the number of replies obtained by the cultural institution. Eventually, the DataVaults approach will allow the lightening of the administrative burden to manage privacy issues in the relationships with data owners, since data always remains in the full possession of the individuals.

In addition, a third scenario has been identified, related with the sharing of personal certificates for fiscal services. In the current approach, whenever requiring a fiscal service from some third party, a citizen has to procure some civil certificate in advance, by either going to the City Registry Office in person or getting it using the online service. In both cases, this represents a burden, and, moreover, the city administration becomes aware of the reason why the certificate is requested since the citizen has to declare it (at least to a broad extent) and this is not exactly in line with the GDPR prescription.

The automatic, authorised, and regular downloading by third parties of shared personal certificates, made possible by the DataVaults tools, will reduce burdens for citizens and interested data seekers. In addition, although the certificates are always produced by the administration, the document exchange is carried out in a totally transparent and privacy-based way without the administration being aware of it, thus making the whole process more compliant with the GDPR legislation.

The application scenarios hypothesized in the DataVaults project can offer advantages in the development of many smart city applications, taking into account the fact that DataVaults offers citizens the possibility to maintain full control of their data and to be compensated for the data made available.

For example, if we consider the theme of digital twins experienced by the region of *Flanders* (Section 3.1), the city of *Pilsen* (Section 3.2), the city of *Athens* (Section 3.9), and other administrations such as *Rotterdam* (Section 3.8), this can be effectively integrated with the approach proposed by DataVaults, as it is possible to hypothesize different types of data provided directly by citizens, which can enrich the information framework managed through the digital twin application. Such data may concern, for example:

- habits related to city mobility (recurring routes, means of transport used, places of interest frequented, etc.);
- environmental data collected through specific devices worn by citizens (for example, temperature, air quality, humidity, environmental noise level, etc.);
- cultural interests (events attended, exhibitions visited, museums, etc.);
- recreational and sporting activities.

Such data can be integrated with data coming from GIS, satellite, local sensors, and other sources in order to improve the quality of the digital city model. But citizens will remain fully in possession of such data, as they can decide how and to what extent they wish to make it available, besides getting rewarded for it.

The data referring to habits related to mobility can of course also be useful for enriching specific projects in this sector, such as the one on electric vehicles developed by the city of *Trikala* (Section 3.4) in Greece. In this regard, for example, the tools offered by DataVaults can be used to administer specific questionnaires to collect feedback on the experiments carried out, allowing the citizens who respond to obtain compensation for their support, when providing additional personal data.

Similar considerations can be applied to the experience carried out by the city of Florence: the administration is currently using a wide range of data source to improve decision-making on mobility, and the connection of the IF App to the data sharing model managed by DataVaults could even more stimulate citizens to use the app, since they could get a compensation for the data they share. In addition, specific questionnaire could be pushed to users who could even get rewarded for their replies. The same approach could also improve the policies on tourism currently adopted by the city of Florence, by helping to build more targeted services on the basis of personal cultural interests.

The scenario hypothesised by the municipality of Prato relating to the use of DataVaults tools for sharing personal certificates with third parties can be effectively generalised in all situations in which there is a need to make certifications of various types accessible to third parties that may require them. Citizens can then personally manage their certificates, professional qualifications, and other personal documents, freely deciding which types of subjects to share them with and possibly also obtaining a fee for this.

Finally, the methods of supplying questionnaires managed by the DataVaults tools can be of great use in many applications related to the development of the smart city, where it is necessary to involve citizens in requesting specific information for the creation of certain smart applications or services. Through the DataVaults tools, the administration can design specific questionnaires and deliver them to all citizens or only to specific categories, according to specific requirements, also associating compensation for those who decide to contribute to the survey. Citizens will receive the request to answer the questionnaire through the DataVaults app and will be able to freely decide whether to do so or not, obtaining the established remuneration if they decide positively to share their personal data.

This approach makes it possible to enhance the contribution of citizens, who will thus be able to feel more involved in the planning of urban services; this will improve the ability of the administration to design services and applications for the smart city on the basis of up-to-date information and to implement more effective planning actions.

The major challenge that cities are now facing is probably climate emergency: Prato is also one of the "100 Climate-neutral Cities by 2030 – by and for the Citizens", selected by the European Commission to showcase systemic transformation towards climate neutrality. This is a key commitment for the administration and citizens have a fundamental role to tackle climate emergency; their involvement and contribution is necessary to build really effective policies to mitigate climate change effects.

The tools provided by the DataVaults project can support city administrations to improve their planning capabilities by, for example, using citizens as "gauges" to detect environmental data (air quality, pollution concentration, etc.) through either apps on their smartphone or wearable devices. Collected data will be accessed in a safe and privacy-respectful way and citizens will be encouraged to contribute by getting compensated for the data they provide. The "by and for the Citizens" motto of the 100 climate-neutral cities can then be fully exploited as one of the main pillars in the development of the smart city.

Digital twins exist as being a digital real-time representation of the physical world. DataVaults can provide access to additional data that is shared that has not been gathered in the digital twin – this can enrich what the digital twin already provides – the use and benefit of this would then be articulated in a tangible smart-city context. DataVaults can provide further insights (through a persona) using data available to DataVaults (or other sources in theory) and those personas can be used alongside digital twins again to enrich the information available to decision-makers, planners, service providers, etc., and again this benefit would have to be articulated in a smart-city context.

4.3 Piraeus's Use of Personal Data

The city of Piraeus has made several steps in the past five years towards becoming an intelligent city. As part of a larger integrated territorial investment program, the amount of approximately 6M€ was directed solely to digital actions. The aim was to make the city more attractive both to citizens and visitors, offering amongst others novel ICT services. Through this funding opportunity and based on the municipality's digital strategy, several areas of interest were touched, as cultural heritage (portal and digitisation of the city's records, library and art gallery), the city's operation monitoring (platform), information gathering through IoT sensors (climate, pollution, waste, energy, illegal parking, air quality, and more), and the destination management system (DMS platform) for tourism. More specifically for the DMS and the city's operation monitoring platform, as the main tools used by the decision-makers in the municipal authority, the amount and quality of information gathered is the crucial element of their success, as the decisions of the city planners are going to be only as good as the information they were based on.

Through DataVaults, the municipality of Piraeus aspires to renovate the way it collects, manages, and analyses the aforementioned information, combining the existing information with citizen's/visitors' personal data and embracing novel engagement practices. The DataVaults app will allow citizens, tourists, and targeted stakeholders to provide more and richer data to the city, entailing more personalised services, always in a trusted and secure manner, respecting their privacy. The data that the municipality aims to access through the DataVaults platform, in combination with existing data, will allow the city to build better analytic reports and identify solutions that could help the local economy, strengthen retail services, improve mobility, promote the local cultural heritage, and meet the touristic engagement goals of the recently established destination management organisation (DMO).

Three different scenarios have been identified and pursued as part of the municipality's participation in the DataVaults project. For each one, we will try to explain the reasoning behind our choice and present the problem and the respective DataVaults solution.

Smart mobility services for individuals:
Piraeus is one of the largest Greek urban centres and one of the most important European ports. In Piraeus, as part of the Athenian urban complex, there are plenty of transportation means in service, i.e., buses, trolleys, suburban railway, tram, train, and underground metro while it is also connected to the central railway network of the country. The combination of sea and ground transport makes Piraeus one of the largest mobility hubs in Greece.

Given the heavy traffic in the city's main road network (40,350 daily movements) and the city characteristics, a major challenge for the local authorities is to take decisions and have accurate information regarding urban mobility. In this context, the municipality of Piraeus is currently finalising its sustainable urban mobility plan (SUMP) and aims, as a next step, to take action within the city in order to reduce traffic, promote public transport, improve air quality, reduce road accidents and safety risks, improve conditions for cyclists and pedestrians, improve mobility for persons with disabilities, and encourage e-mobility.

In the direction of informed decision-making regarding mobility, a small-scale pilot regarding the mobility plan in times of events around Karaiskaki and Peace and Friendship (SEF) Stadiums and the Olympiacos home football and basketball grounds, respectively, is foreseen through the use of personal information gathered from the DataVaults personal data marketplace.

The scenario engages both Olympiacos as a partner in the DataVaults project and the municipality of Piraeus, in order to acquire data shared by the interested citizens, as well as by the Olympiacos members. These data include location (point of origin, route, parking spot, etc.), access time, parking time, and type of transport (bus, train, car, etc.), when someone is accessing the sport venues at the time of an event. The information gathered, along with information already available to the city planners (Municipal Police Department and Programming and Development Department), can then be effectively used to better schedule the mobility strategy and the relevant services around the area of interest.

Empowering local entrepreneurship:
Piraeus is an international cruise centre (over 12 million travellers in 2018) and the link of mainland Greece with the islands of the Aegean Sea and a main commercial hub connecting Europe, Asia, and Africa, providing services to ships and cargo of any type and size, while offering unique advantages because of its strategic position and infrastructure. The high number or travellers, either cruise passengers or commuters to the Greek islands, makes tourism a key factor for the local economy.

The municipality of Piraeus, through the destination management organisation (DMO), aspires to form a concrete plan to engage travellers with the local market. The actions considered include discounts and offerings from the local shop owners in agreement with the Local Traders Association, open markets, festivals (e.g., Piraeus Sea Days), promotion of the local cuisine, and more. The objectives are to (a) motivate cruise passengers to disembark and experience the city and (b) engage commuters with the Piraeus market and local entertainment facilities. DataVaults personal data marketplace offers an invaluable and otherwise unavailable amount of information aiming in the decision-making process. Users will be requested to share their consumer behaviour and preferences, in exchange for discounts on tickets for municipal entertainment venues, or even on specific local shops. Types of data requested from citizens and visitors include commercial preferences and needs, arrival and departure information (for tourists and visitors), location information (routes), and frequency of coming to the city of Piraeus (visitors and commuters). The aforementioned approach provides a dual benefit on the local economy as, on one hand, it helps gather the necessary information and, on the other, it initiates through the compensation mechanism and engagement between the visitors and the local market.

A secondary objective of this scenario is to also invite local entrepreneurship associations (i.e., the Piraeus Traders Association) and other interested stakeholders to join the platform either directly as data seekers or acting as second-tier data seekers, to test aspects of value generation and sharing with entities not directly using personal data but that access the derivatives of the latter.

The "empowering local entrepreneurship" scenario also aligns with the on-going activities of the municipality about the city's Open Trade Centre, associated with the improvement of the local economy through restructuring of the market infrastructures and the deployment of smart applications.

Services for personalised cultural and touristic experiences:
In the city of Piraeus, it is evident that the city has a lot to offer in a variety of areas, such as cultural heritage, entertainment, gastronomy, history, music, and sports. The "services for personalised cultural and touristic experiences" scenario aims to tailor people to the offered experiences. The municipality of Piraeus through the destination management organisation (DMO) once again aims to form a strategy in order to promote the experiences Piraeus has to offer, through actions and events within the city. The Sea Days festival held every year between the end of May and early June hosts more than 100 different events, targeting locals and tourists of different age and interests. The overall strategy and the Sea Days program are planned according to the existing information and previous experience. DataVaults can be an additional source of information, where personal cultural and touristic preferences are shared in exchange for discounts on municipal cultural venues. This information can assist the decision-making process since the DMO gets to know the city's audience and receives invaluable feedback on the events happening in Piraeus. Users will be requested to share their cultural, entertainment, and food preferences, arrival and departure information (for tourists and visitors), location information (routes), and frequency of coming to the city of Piraeus (visitors and commuters). The aforementioned approach provides a dual benefit for the city as, on one hand, it helps gather the necessary information and, on the other, it initiates through the compensation mechanism and engagement between the tourists and the municipal cultural venues. This approach provides a dual benefit also for the user as, besides the compensation scheme, the user may receive suggestions from the DMO for a personalised cultural and touristic experience in the city of Piraeus, based on his individual profile.

4.4 Olimpiacos: Interaction with the Fan-Base

In this scenario, fans and members of the club are encouraged to collect their personal data, understand what it comprises, and manage and share them through the DataVaults platform. For the fans and members, awareness on how the type and quantity of personal data relates to compensation offered from the club shall encourage the former to push forward sharing and keeping this data up to date.

For the club, this will help build a stronger relationship and interaction with the fans and members and understand their needs and offer better services, as well as incentives to share more of their personal data. The club will also be able to respond to the requests that may come from different organisations, such as sponsors, NGOs, sports federations, local authorities, etc., who want to run a campaign or host an event for the club members, fans, and athletes and to find new sponsors and partnerships.

The main goal of this scenario is to collect information related to social media activity and the preferences and likes of individuals who are already fans and members of the club. A secondary goal would be to collect location data related to the position of the individual, for example, if they were present at the stadium during the time the club's team is playing. Collecting and analysing such data could assist the club to combine personal data already existing in its systems with social media and location activity in order to engage and interact better with the fans and members. Further, such information could help reorganise the marketing plan (new market segmentation, marketing campaigns for specific target groups, finding specific sponsors, etc.), attract new sponsors based on personas (collective profiles) extracted from DataVaults, improve the services offered to the fans and members, and, finally, increase the revenues and online presence of the club.

All this, in a transparent exchange between the individual and the club, respecting and enhancing the privacy of individuals and directly compensating those who wish to be part of this incentive.[2]

During the final phase of the project's development, there is sufficient data in the DataVaults cloud platform to elaborate the personas (collective profiles) of, for example, the average season fan or the high-income fan who attends the most prestigious matches. Then, during this phase, the club enables the compensation mechanisms (e.g., in the format of perks) for the members and fans who take part in the initiative. Finally, during this phase, other data seekers will be invited to get on-board; a prominent example of this could be club sponsors who are interested to search through the personal and online activity data of members and fans of the club, in order to offer better customised and targeted products and services. The objectives being:

- to make the branded version of the DataVaults personal app available to a larger group (e.g., upload it to the prominent online app stores or directly invite individuals);

[2] "DataVaults. D6.2-Pilot Scenarios and Implementation Plan," *DataVaults Consortium*. https://www.datavaults.eu/wp-content/uploads/2021/07/DataVaults-D6.2-Pilot-Scenarios-and-Implementation-Plan-v1.00.pdf (accessed Jul. 25, 2022).

- to acquire advanced analytics from the DataVaults cloud platform;
- to have stakeholders from the club inspect the proposed personas from the DataVaults cloud platform and verify their accuracy using their experience;
- to activate the sharing compensation mechanisms and link it with perks for sharing data from the DataVaults app;
- invite club sponsors to use the DataVaults cloud platform as a data seeker.

4.5 Olimpiacos: Athletes Sports and Activity Data Sharing

The second application for Olimpiacos is in relation to their athletes' sports and activity data sharing. The club has 18 different sport departments at the competitive level and the academies which comprise a large base of 2000 professional and young athletes including important contact and medical details. In this scenario, athletes of the club are encouraged to share their personal data, athletic activity data, and ergometric and medical examination data through the DataVaults platform.

For the athletes, awareness on the type and quantity of personal data, athletic activity data, and ergometric and medical examination data belonging to one individual combined with the offering of the appropriate medical and sport equipment by the club and coupled with incentives offered from data seekers to share such data with the platform shall encourage the athletes to share not only already collected but also future athletic activity and medical data.

For the club, this will help with the better management of the sport departments, the planning of the training sections and team tactics, the opportunity to know what specific physical skills are needed when looking into players' transfers, and finally to cover the athletes' expectations by offering the appropriate medical and sport equipment.

The main goal of this scenario is to collect especially athletic activity data and ergometric and medical examination data of athletes who already belong to the club. Collecting and analysing such data could assist the club since the current portal, on the one hand, includes name, surname, address, telephone number, e-mail, date of joining the club, sports (e.g., tennis, sailing, etc.), and on the other hand, it does not allow the entry of more complicated data such as the results of the ergometric and medical examinations. The athlete's base should be also complemented by statistical reports regarding the performance of the athletes in training and during the matches. In those areas, DataVaults could cover the club's need for accessing more complicated information and offering better analytics. All this, in a transparent exchange between the athlete and the club, respecting and enhancing the privacy of individuals. The objectives include:

- making the branded version of the DataVaults personal app available to everyone;
- acquire analytics from the DataVaults cloud platform;
- use DataVaults cloud platform as the main channel to collect athletic activity and ergometric and medical examination data from the athletes who take part in the initiative;
- invite club sponsors to use the DataVaults cloud platform as a data seeker.

The final phases include the introduction of the App to the club's athlete's community and increasing the amount of data available, covering athletic activity and ergometric and medical examination data from the athletes. Other data seekers are on-board; a prominent example of this could be athletic equipment companies who are interested to search through the athletic activity data of athletes of the club, in order to offer better customised and targeted products and services (e.g., depending on what sport each athlete does).

The types of compensation considered in return for the personal data being made available include:

- bonuses (discount coupons) on new official shirts of the multisport club;
- benefits directly related to the activities of the sponsors;
- discount coupons from our club and from other partners.

The expected impact for Olimpiacos from this demonstrator is to be seen in the following areas:

- Effective management of members' and fans' data: Olympiacos has a large basc of members and fans which requires a better data management including collection, mining processing, security, and cryptography. It is very important for the club to secure the personal data protection in the best way to ensure the trust of the individuals.
- As the club develops a large range of innovative and high-quality marketing campaigns, it should emphasise on the collection of valuable insights without violating privacy principles. This database is also

beneficial for the planning and the implementation of segmented and targeted marketing campaigns, the improvement of fan engagement, and finding sponsors. For the club, it would be extremely beneficial to collect information such as user preferences, activity, and mobility data in the future and focus on the real preferences of the members and fans and improve the services and the fan engagement to ensure their loyalty.

- Number of registered club members: Olimpiacos SFP has a large base of 130,000 registered members and 60,000 registered fans. This base is growing rapidly which makes the club the largest sport one in Greece and one of the most popular worldwide. Using the customer relationship management system, the club has built a strong relationship with the member–fan base which seems to be an important issue for the sport successes.
- The data included in the base of members are compulsory to organise and coordinate procedures such as the organisation and the participation of the members in the club General Assembly and the right to vote in the administrative elections and take part in the decision-making.
- Along with the number of active club members and those who have participated in the club's advertised activities and the number of active fans in the club's activities: the number of fans who have followed the club's games in more than one sports department. This base and the satisfaction of fans are very important issues for the club as they are the most important source of revenues.
- Effective management of over 2000 athletes' sports and activity data: Olympiacos has a large base of athletes which requires a better data management including collection, mining processing, security, and cryptography.
- The athletes' base requires better management of the results of the ergometric and medical examinations and statistical reports regarding their performances in the training and the matches. This base also includes personal information, demographics, personal governmental data, and sport exercise/activity one. Better management is very crucial for the club to adapt and plan the training sections and the team tactics, make players' transfers with specific physical skills, and cover the athletes' expectations offering the appropriate medical and sport equipment.
- Revenue from sponsorships: the club organises meetings with the sponsors on a regular basis to discuss the progress of the sponsorship deals

and control the sponsor satisfaction. These meetings could include a presentation of the platform and its utility to motivate the sponsors to join it.

- Stakeholder trust: the club will encourage the members, fans, and athletes to capture, witness, and manage their personal data through this platform and undertake the decision to share them, responding to the requests that may come from different organisations. The club will make clear that the platform ensures the data protection following modern encryption techniques to guarantee privacy and total information security. The club will inform the stakeholders about the possibilities that the platform offers such as full or partial data completeness, encrypted or unencrypted data security, data privacy, etc.

The Olimpiacos use of a citizen's personal data can be shown as adding value to the practical uses of data outlined in Chapter 3, where they reflect tackling issues around mobility. For example, through their relationship with their fan base, travelling arrangements for when the supporters are attending the stadium can be shared with *Piraeus municipality* (Section 3.12) to help their strategic planning, particularly on match days. The immediate outcome of this interaction will result in a decrease in the time required to reach the sports venue and for those travelling by car, a decrease in the time it takes to park in the vicinity of the stadium.

Similar synergies occur with the Tampere (Section 3.6) example described in that chapter. Utilising smart technology (sensors, data analytics, and situation-awareness tools) enables the city to integrate big event areas in the city centre with the surrounding city infrastructure and to support collaboration between event organisers, safety authorities, and third-party actors.

The main sources of data come from city cameras collecting video footage from the city centre. The footage is streamed on a real-time basis to Insta Blue Aware (IBA) situation awareness platform. IBA is a web-based tool providing a joint view to safety authorities, event organisers and event safety services, representatives of other safety operators, as well as to the city safety management team. Based on the information provided by the camera stream, all actors safeguarding event safety and security can create operating models or act accordingly.

By enhancing the supply of data by including citizen's personal data, the result would be better support for collaboration between authorities and non-authorities and for building and sharing of the "big picture" and situation awareness.

With regard to the application concerning the personal data of the athletes at Olimpiacos, it is an answer to a controversy in the world of football in the UK. Hundreds of footballers have threatened legal action against the data collection industry, which could change how information is handled. 850 players want compensation for the trading of their performance data over the past six years. They also want an annual fee from the companies for any future use. "Letters before action" have been sent to 17 big firms, alleging data misuse. While receiving a fee for the use of their data might not have much impact on the high earners of the Premier League, it is felt strongly that those lower down the pyramid, in both the men's and women's game, would see tangible benefits. Russell Slade, a group leader, pointed out that "On one player, and I'm not talking about a Premier League player or even a Championship player, there was some 7,000 pieces of information on that one individual player at a lower league football club".

"There are companies that are taking that data and processing that data without the individual consent of that player". He drew into question other sports' similar behaviour. "A big part of our journey has been looking at that ecosystem and plotting out where that data starts, who's processing it, where it finishes and that's a real global thing. It's making football - and all sports - aware of the implications and what needs to change".[3]

This is in stark contrast to the approach being taken by Olimpiacos.

4.6 Andaman7 Health Application

Andaman7 is a patient-centric mobile tool for collaborative management of medical records between health professionals, patients, authorised relatives of such patients (parents or spouse, for example) and connected devices. Exchange and synchronisation of information are peer-to-peer, based on original concepts maximising security and confidentiality.

With the Andaman7 app, patients can:

- feed their medical records by importing health data: the platform captures data from electronic medical records, sensors, wearable, and quality of life questionnaire data.
- access their data in a structured, secure, and standardised way and consult them anytime, anywhere;

[3] "Professional footballers threaten data firms with GDPR legal action - BBC News." https://www.bbc.com/news/uk-wales-58873132 (accessed Jul. 25, 2022).

- share their file with authorised third parties: general practitioners, specialists, or family members.

It is more than "a simple mobile app". It is an advanced health IT software combining mobile applications, server processing, smart peers, and a full HIP – health interoperability platform – for the easy and efficient distribution of data, while preserving the privacy of patients and the security of the overall system.

The solution touches on a universal problem: human health and our business model are based on a wide distribution of the application in "Freemium". As such, revenues come from paid advertising and add-ons (including clinical studies apps).

The application is currently well differentiated and complementary to existing solutions and not a competing solution to electronic health records. There are more than 22.000 users registered on the Andaman7 platform coming from 35 countries. The platform is available in 22 languages, and 80% of the users are from Europe. Andaman7 is also connected to some hospitals, laboratories, and pharmaceutical companies to either send personal data to the patient or gather information from them (as in clinical trials).

In the DataVaults project, current users of Andaman7 will be able to connect to DataVaults to store all or part of their health data. This storage can be used as a backup to retrieve data when lost. This can also be used by third parties in the health sector (e.g., clinical trial, research, etc.).

The main objective for Andaman7 is to improve the general attractiveness of the application for both users and partners and thus increase the number of users and the number of services being provided by the integration of new features and creating new partnerships. This entailed developing the backup of Andaman7 content through the DataVaults platform as this is a sought-after feature by a lot of users to increase confidence towards the application and especially the storage of their data.

And, second, to develop and run a fictitious clinical study to show the potential of the DataVaults platform associated with Andaman7 and communicate to potential partners about the results. This will be measured through analytics tools integrated in the Andaman7 platform. As the project matured, the final phase for the clinical trial was based on the retrieval of such data and what we can get from the platform to help running real clinical trials and/or improve health in general. The intention is to exploit as much as possible all features provided by the DataVaults platform to provide meaningful data to clinical users, document them, analyse the results, and present it to potential data seekers. This entails the following:

- retrieval of clinical study data from a data seeker perspective with compensations mechanism with analytics and statistics data retrieved;

- documentation of results obtained from this demonstration study;
- presentation of the solution and obtained results to Andaman7 partners for on-boarding.

A second scenario related to data collection whilst improving Andaman 7. Current users of Andaman7 will be able to connect to DataVaults to collect their data (coming from various sources) and store them in Andaman7 on their smartphone. This will make the data available to patients for reviewing, learning, and using in other setups (e.g., share additional data with their doctors, hospital, etc.). Data will mostly be raw personal data but also aggregated data (e.g., result of a clinical trial, comparison to a specific group, etc.). Further objectives were added; of increasing the number of supported data types by 20%, of increasing the quantity of data available in each category and/or also increasing the number of categories.

The interest of the final user is the main goal here as getting more meaningful data from various sources to complete their record should attract more users. To achieve those goals, in this scenario, we will:

- complete the health record of patients with health data available on the DataVaults platform;
- compute some statistics based on data of the patient or other users of the DataVaults and integrate them in the health record of patients as indication to bring some intelligence to the storage of data.

This will be measured through analytics tools integrated in Andaman7.

This final phase will bring more value to the users by trying to match them with people with similarities in their medical profile, retrieve statistics/analytics data from this community, and display it to the user so that he can compare his results to others. The identification of data that will be retrieved and displayed can be sensitive as it should not be considered as medical advice. This requires:

- identification of criteria to find people that have similarities in their medical records and corresponding statistics/analytics data that could be interesting;
- development of connectors to retrieve statistics/analytics data from people that have similarities;
- retrieval, conversion, and insertion of data into Andaman7 smartphone app.

In the medical field, data seekers place great emphasis on the amount of collected data, the quality of data, and the persona of the data owner. Through the

DataVaults platform, they will be able to find and gather data more easily – data from a specific target group while reducing R&D costs. As all pieces of data are stored in the same place, they can get a huge amount of data (raw or computed) easily. Quality can also be easily checked because the owner of the data is known and can be verified by the platform.

Chapter 8 further discusses the Andaman7 involvement in this area, with a discussion of the InteropEHRate project and interoperability within the health sector.

4.7 Smart City Graz

The My Smart City Graz district is located in the vicinity of the Helmut-List-Halle. This former industrial area is located right in the centre of Graz west of the Graz main train station. The former commercial area is an important building land reserve that is being developed into a sustainable place to live and work with a high quality of life. The aim is a sustainable energy supply and resource conservation at all levels. In addition, new living space and a high-quality public space are to be created. It combines New European Bauhaus ideas with smart city progress.

At the heart of this initiative is the "Science Tower", which is a 60-m-high research tower to be erected north of the Helmut-List-Halle. The tower is an energy self-sufficient building that will house offices for Graz-based companies from the "Green Sciences" sector. The area above the offices is designed as a roof garden, with a meeting room in the middle. A roof garden was created on the upper floor.

This 60-m-high tower is located north of the Helmut-List-Halle and not only houses science and research but is itself also a research object for new building technologies. Up to the 12th floor, there are offices (approximately 44-m high), in which primarily Graz companies from the "Green Sciences" sector are rented. Together with the surrounding companies and universities, they form a "science cluster" – i.e., a combination of know-how, technologies, and ideas for an ecological and sustainable future. The area above the offices was designed as a roof garden where fruit and vegetables are grown.

Regarding the use of personal data, these companies and universities within this science cluster will be able to utilise the work of the KRAKEN project.[4] The KRAKEN project application provides an exemplar of how

[4] "Health pilot - Kraken." https://www.krakenh2020.eu/sites/kraken/files/public/content-files/2020/Kraken%20rollup-pilots.pdf (accessed Jul. 25, 2022).

personal data can enhance operations within this smart tower. The Technical University of Graz has led the development of the KRAKEN application which helps to release personal data for value creating use within the academic sphere.

Data from academic students (e.g., graduation certificates, course grades, enrolment status, etc.) is protected with privacy-preserving cryptography, authenticated using self-sovereign identity solutions, and shared in a marketplace. This is of benefit to students, recruiters, universities, and for companies in the following ways.

- For students: sharing selected academic data for job applications.
- For recruiters: identifying the top job candidates by analysing student datasets.
- For universities: exchanging students' data while ensuring their privacy.
- For companies: finding new talent more easily through greater access to student data.

5

The Local Data Economy

Shaun Topham

Lead, EU Smart Cities Marketplace, "Citizen's Control of Data" Initiative
Email: shaun.topham@eu-forum.org

Abstract

In this chapter, we will very briefly raise the topic of how the local data economy which exists in a smart city can benefit from having access to a citizen's personal data, as part of the wider ecosystem, which, if successfully created in a city, should produce data at a scale of real value to local companies and to SMEs in particular, thus enhancing operations already being carried out in the local data economy and stimulating new ones.

An underlying objective of this book has been to try to indicate which developments are in the pipeline and so should be taken into account when planning the evolution of urban data platforms and growing the local data economy, enhanced with personal data. This chapter will provide examples of ongoing work to stimulate the data economy – and hence from a smart city perspective, the local data economy.

5.1 Introduction

Chapter 4 has covered how by making personal data from the citizens available in a trusted way, it can augment the smart city applications which were highlighted in Chapter 3, as well as generally improving service delivery, whilst engaging citizens in the decision-making process towards the goal of achieving climate – neutral cities by 2030.

The following chapters will cover a range of topics tackling other issues faced, before personal data can be fully utilised. Here we will very briefly raise the topic of how the local data economy, which exists in a smart city, can benefit from having access to a citizen's personal data,

DOI: 10.1201/9781003399384-6

as part of the wider ecosystem, which, if successfully created in a city, should produce data at a scale of real value to local companies and to small and medium enterprises (SMEs) in particular, thus enhancing operations already being carried out in the local data economy and stimulating new ones.

It has been said that an SME operating in the data economy finds it easier to raise finance than it does to access the data it requires. Creating an ecosystem to access personal data should encourage wider sharing of data and, therefore, provide a stimulus to SMEs.

If the EU's ambition of increasing its share of the global data economy commensurate with the size of its economy is to be reached, here is a driver. And the required shift of power from the huge non-European corporations in the direction of a European model of the data economy will rely considerably on SMEs meeting the challenge, aided in part by the EU funded projects which are working in this field.

The EU data market is forecasted to reach 85 billion euro with accumulative growth of 7% between 2020 and 2025 as European enterprises accelerate their digital transformation.[1]

Data sharing and data interoperability are still at their infancy; few data markets for sharing industrial data exist. In a recent survey, more than 40% of the SMEs interviewed claim that they had problems in acquiring data from other companies. The diffusion of platforms for data sharing and the availability of interoperable datasets is one of the key success factors which may help to drive the European data economy and industrial transformation.[2]

The smashHit project sums up the potential value of data to a local economy:

"This enormous amount of data offered by various data sharing platforms represent:

- A massive information resource to create new value, allowing the improvement of existing services or the establishment of diverse new innovative services, by combining data streams from various sources.

---

[1] "The European data market study update - Shaping Europe's digital future." https://digital-strategy.ec.europa.eu/en/library/european-data-market-study-update (accessed Jul. 25, 2022).

[2] "SME panel consultation - B2B Data Sharing - Shaping Europe's digital future." https://digital-strategy.ec.europa.eu/en/library/sme-panel-consultation-b2b-data-sharing (accessed Jul. 25, 2022).

- A major big data-driven business potential, not only for the manufacturers of Cyber Physical Products (CPP), but in particular also for cross-sectorial industries and various organisations with interdisciplinary applications".[3]

An underlying objective of this book has been to try to indicate which developments are in the pipeline and so should be taken into account when planning the evolution of urban data platforms and growing the local data economy, enhanced with personal data. This chapter will provide examples of ongoing work to stimulate the data economy and, hence, from a smart city perspective, the local data economy.

Chapter 2 covered a range of projects that are contributing to moving forward this agenda, as "Innovation Actions for setting up and operating platforms for secure and controlled sharing of "closed data" (proprietary and/or personal data)".[4]

Whilst the focus has been on projects working towards making personal data more accessible, other sister projects have been facing the challenge of helping the local data economy by providing:

- privacy metrics that are easy to understand for data subjects and contribute to the economic value of data by allowing privacy-preserving integration of independently developed data sources;
- industrial data platforms that shall enable and facilitate trusted and secure sharing and trading of proprietary/commercial data assets with automated and robust controls on compliance (including automated contracting) of legal rights and fair remuneration of data owners;
- the need to link to and bring in industrial data providers that will populate the platforms, with conditions of use and practical arrangements of data sharing being regulated.

We will look in more detail at what the projects i3-MARKET, AURORAL, and smashHIT are achieving as they have focused more on the economic aspects, but we should also point to other projects contributing to improving the local data economy.

---

[3] https://www.smashhit.eu/wp-content/uploads/2021/03/smashHit_D1.3_Public_Innovation_Concept_v100.pdf

[4] "Supporting the emergence of data markets and the data economy - Programme H2020," *CORDIS - European Commission.* https://cordis.europa.eu/programme/id/H2020_ICT-13-2018-2019 (accessed Jul. 25, 2022).

These include:

- MOSAICrOWN[5] which is providing "effective and deployable solutions allowing data owners to maintain control on the data sharing process, enabling selective and sanitised disclosure providing for efficient and scalable privacy-aware collaborative computations".
- OpertusMundi[6] which is delivering "a trusted, secure, and highly scalable pan-European industrial geospatial data market, which will act as a single-point for the streamlined and trusted discovery, sharing, trading, remuneration, and use of proprietary and commercial geospatial data assets".
- SmashHit[7] which, in addition to assuring trusted and secure sharing of data streams incorporating personal data within a smart city context, also considers the industrial platforms that are needed to build sectorial and cross-sectorial services.

There are a variety of other projects which are contributing more at the European and specific sectorial levels.

Two projects highlighted in Chapter 2, AURORAL[8] and EUHubs4Data,[9] have emerged from the industrial leadership programme – leadership in enabling and industrial technologies – information and communication technologies (ICT).[10]

The programme has the aim of mastering increasingly complex and multidisciplinary technology and business chains in ICT, with partnering, risk-sharing, and mobilisation of critical mass across the union being needed.

The European DIGITAL SME Alliance plays a leading role in both the i3-MARKET and AURORAL projects, in order to assist in the development of its members and provide more opportunities for them, such as those which should materialise with the availability of more personal data to the existing

[5] "MOSAICrOWN – Multi-Owner data Sharing for Analytics and Integration respecting Confidentiality and OWNer control." https://mosaicrown.eu/ (accessed Jul. 25, 2022).

[6] "Opertus Mundi." https://www.opertusmundi.eu/ (accessed Jul. 25, 2022).

[7] "Smash Hit." https://smashhit.eu/ (accessed Jul. 25, 2022).

[8] "AURORAL." https://www.auroral.eu/#/ (accessed Jul. 22, 2022).

[9] "EUHubs4Data - European Federation Of Data Driven Innovation Hubs." https://euhubs4data.eu/ (accessed Jul. 25, 2022).

[10] "INDUSTRIAL LEADERSHIP - Leadership in enabling and industrial technologies - Information and Communication Technologies (ICT) - Programme H2020 ," *CORDIS - European Commission*. https://cordis.europa.eu/programme/id/H2020-EU.2.1.1. (accessed Jul. 25, 2022).

and evolving ecosystems within the data economy. The Alliance, the largest in Europe with more than 45,000 digital SMEs represented, is the joint effort of 30 national and regional SME associations from EU member states and neighbouring countries to put digital SMEs at the centre of the EU agenda.[11]

We will look in turn at the contribution these two projects are making towards growing the local data economy.

5.2 i3-MARKET

The main objective of the i3-MARKET project[12] was to develop the missing building blocks standing in the way of having an "Intelligent, Interoperable, Integrative and deployable open source MARKETplace backplane with trusted and secure software tools for incentivising the industry data economy".

This is in the form of a software framework called the "i3-MARKET backplane" for data providers and consumers, and, thus, it incentivises and enables the creation of a more trusted European data market economy. It addresses the interoperability and integration challenges for trading data assets across independent stakeholders by means of secured transactions based on data annotation (semantics), as well as a trusted data trading platform, and will provide a network of decentralised and economy-driven and scalable data repositories that can be extensible for enabling the deployment of intelligent industrial data services fostering innovation and business opportunities.

This backplane is now live and can be accessed at their webpage.[13]

It represents the culmination of the first phase of the project: the backplane is the results of the reference architecture, the mapping of the security, privacy and trust requirements, and will now be tested in three different industrial settings. It is providing the support tools to promote a joint ecosystem for unified European data marketplaces and data spaces, which can address the diverse needs that the demand side for data markets and spaces creates, and as a service provides a one-size-fits-all solution. It is based on open source technology and architecture and exists to provide secure services to the expanding data market and data economy. The backplane will fill a gap in the current data economy, as digital services are suffering from a lack

[11] "European DIGITAL SME Alliance." https://www.digitalsme.eu/ (accessed Jul. 25, 2022).

[12] "i3-Market." https://www.i3-market.eu/about/#project (accessed Jul. 22, 2022).

[13] "i3-MARKET Backplane Goes Live - i3 Market." https://www.i3-market.eu/2021/05/27/i3-market-backplane-goes-live/ (accessed Jul. 25, 2022).

of support to ensure reliability and robustness, which it can offer through a federated, fully distributed environment, and thus providing an incentive for local data economies to grow and for future platforms to evolve.

It will allow local SMEs (as well as other stakeholders) to share, trade (monetise) their data assets, as well as to use and buy it from the others. Thus, it will incentivise and enable the creation of a more trusted European data market.

The piloting has taken place focusing on three sectors[14]:

- the automotive sector, bringing data supply and demand together and providing the possibility of efficient matchmaking through an open and brand-independent platform;
- the intelligent manufacturing sector, with SIEMENS offering their data on their products, manufacturing and supply processes through a single dataspace or marketplace;
- the healthcare and wellbeing sector, with IBM providing various elderly care solutions and products, which will allow this pilot to leverage IoT technologies to address patients' and relatives' needs.

5.3 AURORAL

We will now turn to AURORAL[15] which stands for "Architecture for Unified Regional and Open digital ecosystems for Smart Communities and Rural Areas Large scale application".

Whilst the AURORAL project is focusing on smaller communities, its overall ambitions and focus will also be of benefit to all communities, regardless of size. As well as the technology being produced, lessons learned in governance of such evolving eco-systems will be of value.

AURORAL focuses on increasing connectivity and delivering a digital environment of smart objects, interoperable services platforms able to trigger dynamic rural ecosystems of innovation chains, applications, and services. Thus, AURORAL contributes to increasing economic growth and creating jobs and to tackling significant societal challenges and contributes to overcoming the digital divide between rural and urban areas and to developing the potential offered by increased connectivity and digitisation of rural areas.

---

[14] "i3-MARKET – The missing link between Europe's data marketplaces." http://open-source.i3-market.eu/ (accessed Jul. 25, 2022).

[15] "AURORAL." https://www.auroral.eu/#/ (accessed Jul. 25, 2022).

AURORAL digital environment is demonstrated by cost-efficient and flexible cross-domain applications through large-scale pilots.

Piloting is taking place in: Alentejo (Portugal), Southern Burgenland (Austria), Hålogaland (Norway), Västerbotten (Sweden), Lapland (Finland), Piedmont (Italy), and Penedès (Catalonia, Spain).

The project is focusing on five topics – tourism, mobility, energy, health, and agriculture.

It builds on an open, API-based, interoperable and federated Internet of Things (IoT) architecture and includes a reference implementation supporting flexible integration of heterogeneous services, bridging the interoperability gap of the smart object platforms and creating markets for services.

The AURORAL reference architecture is devised to provide the means to several different diverse local communities from different domains, in order to be part of a dynamic rural ecosystem to:

- connect and share data collected locally through a secure and privacy-preserving framework;
- engage external technology and application providers in exploiting their data by offering advanced horizontal services to process and create value out of these data;
- participate in new dynamic online marketplaces as commodities services and online platform operators;
- implement easily in an interoperable way and based on open application interfaces based on open standards.

AURORAL is built on top of a previous H2020 project, VICINITY.[16] It is extending VICINITY's core functionalities, in order to promote and facilitate a smooth integration of different entities and services, by the extension of the core platform to a data brokerage middleware, enabling flexible integration of heterogeneous vertical and horizontal tools and services from cross-domain applications.

Thus:

- providing enhanced semantic discovery and interoperability features;
- featuring advanced access control management and end-to-end security, privacy, and trust, preserving data sovereignty;

[16] "Vicinity - Open virtual neighbourhood network to connect IoT infrastructures and smart objects." https://vicinity2020.eu/ (accessed Jul. 25, 2022).

- shifting from a centralised storage to a distributed storage component based on DLT smart contract management, facilitating scalable data access and sharing;
- integrating Elliot Cloud middleware to further facilitate the use of core FIWARE components;
- enabling easy middleware integration of external tools/devices/services by open APIs based on open standards.

The AURORAL reference architecture aims to facilitate the integration of different vertical tools and services from the targeted domains of energy, mobility, tourism, health, and agriculture.

Examples of solutions being produced, whilst aimed at improving rural and smaller community life, but equally appropriate to cities in most cases, include the following.

Tourism: AURORAL bundles all the touristic activities and services from a rural region. The platform connects and integrates many existing data bases and services from different sectors. For example, the connection of the mobility sector to the tourism sector by integrating shuttle services and taxi services with touristic activities. By using a provided single sign on technique, guests can either book just the tour or can book the tour in combination with different transportation options. Big data analysis in the tourism domain, for example, can bring in the whole guest journey from all event logs for digital touristic systems (e.g., oHA,[17] sensors, or web pages from local companies). For example, processes can be discovered which show the common user behaviour for tourists before booking a massage. This involves use of personal data.

Digitalised health services: Digital care and eHealth solutions for elderly or people with multiple chronic conditions, especially those living in rural areas, provide the following benefits: (a) improving access to health care services in areas with low (or no) availability of relevant services as eHealth tools can enable remote consultations, therapies, and rehabilitation; (b) enhancing care coordination and integration as eHealth solutions can help with collecting, storing, and reporting health data to professionals and to patients via electronic health records (EHRs) and personal health records (PHRs); (c) supporting decision-making by clinicians as decision support systems (DSSs) can link available clinical evidence on appropriate treatments and best practice for the patients profile improving the quality of care

[17] This is a platform developed by LuxActive https://www.luxactive.com/en/

provided; (d) enabling monitoring, risk analysis, and proactive intervention since an information system for risk stratification can monitor and predict health risks in a population as well as indicating recommended strategies for prevention, monitoring, and treatment.

Digitalised energy: Nowadays, the provision of energy in a rural context is not about bringing electricity to consumers but to empower them by providing the means for combining consumption with production of electricity through renewable sources, local temporary storage, and demand management. All of these just require a single communication channel and a managing platform which can be common to other sectors. It is based on an open source platform[18] that enables the integration of devices at the edge by fully exploiting data available from local and distributed energy resources to build added value services for energy stakeholders. Now, its continuous improvements have led first to develop tools and services to enable transaction management for local energy markets as for the decentralised management of energy supply and demand and, as a consequence, the management of any kind of goods or services that can be provided by members of a community using a multi-carries and multi-commodities approach (energy, mobility, health or education, agricultural, livestock production, etc.).

The AURORAL project will be enabling the integration of several different marketplace application areas and commodities services, ranging from:

- assisted living as a platform;
- intelligent healthcare and social assistance operations;
- digital transportation services;
- intelligent logistics hub;
- mobility-as-a-service;
- energy trade flexibility service;
- online markets with rural products and services

5.4 The smashHIT Project

The smashHIT project is based on the need to overcome the obstacles caused by wide variety in technical designs and proprietary implementations and

[18] "FUSE - Research & Innovation." https://booklet.atosresearch.eu/fuse (accessed Jul. 25, 2022).

locks down various business opportunities due to all the inconsistent consent and legal rules among all the different existing data-sharing platforms, operators, and bodies involved. In turn, this restricts innovation and results in lost opportunities.

The project will offer a new framework for effective sharing and brokerage of data streaming over diverse platforms, both personal and industrial, and allowing for efficient generation of services by combining the data. The main focus in smashHIT will be on platforms with data streams coming from the usage of products embedded within systems in combination with data from personal and industrial platforms covering smart city data, as well as insurance data platforms. They too point to the fact that providing access for the entire local economic community will become a powerful resource for further innovations. Tools they provide on their "3D city operations platform" – an open data platform – will help every third party to use data and visualisation tools to create their own business. "This will influence society and lead to additional economic benefits for companies and more wellbeing for citizens".[19]

5.5 The smashHIT Methodology

We will return to the fact that all the projects referred to seek to make their results available and with the necessary support tools to encourage take-up. It is worthwhile to draw attention to the approach taken by smashHIT to do this.

Besides the specification and implementation of the smashHit platform solution (smashHit platform, consent certification, data use traceability, etc.) which will be described in Chapter 6, the project has foreseen to develop and provide a set of support material, which serve the different stakeholders of the smashHit value chain (data owners, data providers, data consumers, and data processors) as basic guidelines for participating in the overall smashHit workflow.

A key challenge of the project's methodology is to address a privacy-aware consent and contract management for secure sharing of data from and between diverse platforms, including agreed consent and legal rules among stakeholders (data providers, data owners, and data users).

Therefore, the methodology will provide guidelines on how to use the smashHit platform and the privacy- and security-preserving services to

[19] "What is the smashHit Project? - Smash Hit." https://smashhit.eu/smashhit-press-release/ (accessed Jul. 25, 2022).

integrate and assure data owner and service users consent and compliance with legal rules of all involved stakeholders over diverse platforms. The methodology will provide the method to translate the legal rules of the stakeholders in the semantic model which will be used by smashHit solution to achieve a platform overlapping consent certification.

In this context, one of the key objectives of the support material is to empower smashHit platform stakeholders to:

- easily implement specified and developed procedures and tools;
- to understand offered functions and tools.

The developed support materials will also address organisational, administrative, and contractual measures concerning the interaction of the various stakeholders with the smashHit platform.

To achieve such envisaged methodology concept and to identify required tailored support material to be developed, the specific view and needs for each of the smashHit platform solution stakeholders was taken into account, which results in the following four types of support materials.

Developer guidelines: This type represents specific developer guidelines for industrial smashHit platform users (data providers, data consumers, and data processors), aiming to give these user groups all necessary knowledge on how to connect their systems with the heart of the smashHit platform (smashHit platform).

User guidelines: This type represents specific user guidelines for all types of smashHit solution end-users (data owners, data providers, data consumers, data processors, and smashHit platform administrator), aiming to help users on how to use the platform (e.g., the specific functionalities provided by the platform, such as user authentication, consent/contract certification, data use traceability, automatic contracting, etc.) and gives knowledge about the different functionalities available.

Concept papers: These documents provide easy understandable general descriptions of the key innovative project outcome, like the overall smashHit platform solution.

smashHit platform policy guidelines: As a complementary support material, this document covers legal, privacy, and consent regulations for key processes/activities with respect to the actions/roles of the various stakeholders and their interaction required for the operation of the smashHit platform solution.[20]

[20] https://www.smashhit.eu/wp-content/uploads/2021/03/smashHit_D1.3_Public_Innovation_Concept_v100.pdf

The final set of smashHit support material will be presented in the scope of Deliverable 2.2 and will be made publicly available.

5.6 Conclusion

We saw an example in the previous chapter of how Piraeus is attempting to utilise a citizen's personal data to help support the local economy. In this instance, the personal data for the platform is that of the very large number of travellers, embarking on the ferries and cruise ships which make tourism a significant factor for the local economy. Additional personal data was exchanged for "offers" which motivated travellers to experience the city and engage them in the local attractions and facilities. This approach provides a dual benefit on the local economy, as, on one hand, it helps gather the necessary information, otherwise unavailable, and, on the other, it initiates through the compensation mechanism and engagement between the visitors and the local market. Stakeholders are encouraged to join the platform as data seekers and data providers in their own right, to test aspects of value generation and sharing with entities not directly using personal data but that access the derivatives of the latter. This fits with the overall economic development strategy for improving the local economy through restructuring of the market infrastructures and the deployment of smart applications.

In this chapter, we have touched on a plethora of obstructions to the growth of a local data economy, but also with many solutions being suggested.

Chapter 6, "Technical Solutions", aims to give a simplified overview of the technology currently becoming available.

In this context, the RUGGEDISED project also concluded that[21]:

> "from a technical perspective, a platform should also provide business model support tools to enhance the economies of scope by encouraging new communities (e.g., data-driven start-ups, developers, and established firms) to join the platform ecosystem in order to explore new business opportunities, or to enhance their existing business models".

---

[21] "RUGGEDISED D6.6 Governance, Trust and Smart City Business Models: the Path to Maturity for Urban Data Platforms," *RUGGEDISED Consortium*, Accessed: Jul. 25, 2022. [Online]. Available: https://ruggedised.eu/fileadmin/repository/Publications/RUGGEDISED-D6.6-Governance-Trust-SmartCity_business_Models-EUR-FINAL-2020.11.13.pdf

And that: “Within the RUGGEDISED platform architecture are some key technical capabilities that will enhance the trust in the Urban Data Platform. These are security, privacy, analytics and business model support tools. Security and Privacy are a *sine-qua-non*. Analytics and Business model support tools are a matter of platform maturity. The better the business model support tools, the easier it is for innovators to enhance their existing business models or to create new business models by using the Urban Data Platform. Software Development Kits, APIs, and data marketplaces are all enablers of new business models. The better these tools, the more likely it is to succeed. Such technical tools allow access and interaction with the platform and mediate between the platform and its users. Hence, they play an important role in opening up new business opportunities within a platform ecosystem”.

Chapter 11 will continue this economic theme, looking at the emerging business models within the interlinked smart cities, data, and personal data economies. And subsequent chapters will look at how to overcome those other obstacles that have been identified, including interoperability and standardisations issues and financial, governance, and legal issues.

6

Technical Components

Sotiris Koussouris[1] and Yury Glikman[2]

[1]Suite5, Cyprus
[2]Fraunhofer, Germany
Email: sotiris@suite5.eu; yury.glikman@fokus.fraunhofer.de

Abstract

This chapter covers a selection of the technologies coming out of those projects listed in Chapter 2, which collectively contribute to a citizen being able to safely share their personal data.

The four sections reflect the different general issues being faced and have attempted to reflect the objectives of the European Commission when these projects were granted their funding.

They include:

- data owners and subjects controlling their own data;
- preserving data privacy and data quality simultaneously;
- information delivery on privacy metrics and data content and value;
- data platforms.

Additionally, reference is made to several initiatives which are designed to support this overall process in making it possible for a citizen to share their data securely and fully under their own control.

6.1 Introduction

As an indication of what is being produced by the projects listed in Chapter 2, some of the technologies that they are producing are highlighted, along with references to some of the other supporting initiatives, working within this field.

DOI: 10.1201/9781003399384-7

This chapter covers the technologies that provide the basis for enabling a citizen to be able to share their personal data securely and in a way that they themselves have control. These are a combination of established technologies and those emerging from the raft of projects funded through the Horizon 2020 Research Framework Programme, principally in the ICT-13-2018-2019 Call "Supporting the emergence of data markets and the data economy",[1] which was described in Chapter 2, in relation to why we have focussed upon certain projects. These projects were amongst other things, "Supporting the emergence of data markets and the data economy".[2]

The chapter should be treated more as a "catalogue" to act has a guide as to what the projects covered in Chapter 2, "have been up to", rather than as a learned treatise. It is divided into four sections, which reflect those technologies signposted within the above-mentioned, which, in the words of the call, are:

- "The personal data platforms shall ensure respect of prevailing legislation and allow data subjects and data owners to remain in control of their data and its subsequent use".
- "Solutions should preserve utility for data analysis and allow for the management of privacy/utility trade-offs, metadata privacy, including query privacy".
- "Solutions should also develop privacy metrics that are easy to understand for data subjects and contribute to the economic value of data by allowing privacy-preserving integration of independently developed data sources".
- "Industrial data platforms shall enable and facilitate trusted and secure sharing and trading of proprietary/commercial data assets with automated and robust controls on compliance (including automated contracting) of legal rights and fair remuneration of data owners".

6.2 Data Owners and Subjects Controlling their Own Data

Over the last years, witnessing the advent of various data-driven platforms that exploited in a very effective, sufficient, and profitable manner data belonging to users, a huge societal movement was constructed calling for

[1] https://cordis.europa.eu/programme/id/H2020_ICT-13-2018-2019

[2] "Supporting the emergence of data markets and the data economy. Programme H2020," *CORDIS European Commission.* https://cordis.europa.eu/programme/id/H2020_ICT-13-2018-2019 (accessed Jul. 25, 2022).

new methods and tools, both from the technology as well as the legislative level, for allowing data owners and subjects to regain control of their data and their subsequent use and bring them into the position of the decision-maker to resolute how, when, for which cause, and by whom data shall be used.

In Section 6.2, we explore novel technological approaches that contribute towards this direction.

6.2.1 User personas

User personas have their origins in the marketing world and are used as another way to preserve the privacy of individuals. A persona is a mechanism by which insight from aggregated datasets (groups of individuals) is generated and shared with those who are interested (data seekers in this case).

As an example, we can think of a data seeker who can ask for a Persona on the daily physical activity of people living in Paris. The technical workflow to generate such a persona is as follows:

1. The data that is irrelevant to the query is dropped from the dataset.
2. The persona generator clusters the data using a mean shift algorithm and adds a new column to the dataset and assigns a number showing which cluster each data point is a part of.
3. For each column, mean, mode, standard deviation, and variance are calculated and put into a dictionary.
4. Similarly, a coefficient matrix is produced and stored in this dictionary as well.
5. Then the same statistics are produced for any categorical columns in the dataset.
6. This process is iterated for each cluster in the dataset.

These insights are essentially demographics relating to the statistical properties of the aggregated data. The resulting persona of our example provides insights on the average daily activity of people living in Paris broken down by age and area, the main data points that determine higher and lower levels of activity and how closely these data points are related. This means a data sharer can still get some value from their data without actually letting a data seeker see his/her data, thereby reducing the risk of them being identified.[3]

[3] https://www.datavaults.eu/material/results-and-documentation/

6.2.2 Direct anonymous attestation (DAA)

Direct anonymous attestation (DAA) is a technology based on group signatures for enhancing privacy and trust amongst transacting stakeholders.

From a technological perspective, DAA is a cryptographic protocol that allows a trusted platform module (TPM) to serve as a trust anchor for a host platform it is embedded in. To do so, the TPM chip creates attestations about the state of the host system, e.g., certifying the boot sequence the host is running on. These attestations convince a remote verifier that the platform it is communicating with is running on top of trusted hardware and using the correct software. A main design goal of DAA is that attestations are made in a privacy-preserving manner. That is, the verifier can check that attestations originate from a certified hardware token, but it does not learn anything about the identity of the TPM. Another important feature of DAA is that it supports user-controlled linkability which is steered by a base name "bsn." If a platform uses a fresh or empty base name, the resulting attestations cannot be linked, whereas repeated use of the same base name makes the transactions linkable. A DAA can be seen as a special variant of group signatures with a central issuer controlling membership to the group of certified TPMs, and TPMs are able to sign anonymously on behalf of the group. Instead of the opening capabilities provided in group signatures, DAA controls privacy through the use of base names and user-controlled linkability.

The privacy requirements that are captured by DAA are the ones documented in the ETSI TS 102 941 standard[4]: anonymity (ability of a user to use a resource without disclosing its identity), pseudonymity (ability of a user to use a resource without disclosing its identity while being accountable for that action), unlinkability (ability of a user to make multiple uses of resources without others being able to link them together), and unobservability (ability of a user to use a resource without others being able to observe that the resource is being used). The project DataVaults[5] provides a good example of all the above cases where DAA is used for enabling data owners to both authenticate their platforms in a privacy-preserving manner and also share their data in an anonymous way by leveraging group-based pseudonyms.

---

[4] "ETSI ICT Standards." https://www.etsi.org/standards#Pre-defined%20Collections (accessed Jul. 25, 2022).

[5] https://www.datavaults.eu

6.2.3 Access control policies

One of the key points when it comes to controlling data is to solve the fundamental issue of how citizens can become the actual owners of their data setting the conditions to be fulfilled, in a transparent and automatic way, for external data seekers. Currently, the perception of the citizens about being in charge of controlling with whom their data is shared is very low, making the decision to share not as easy as the modern smart cities would need. The GDPR, as a means to protect their data against abusers, is quite new and barely known by regular people. Moreover, there are not enough incentives (for example, monetisation) to promote data sharing. Nevertheless, more and more data is being generated with a real possibility of using them to improve lots of services.

One possible approach, as the one employed by the DataVaults project, builds on the premise to change this feeling of the citizens based on letting data owners decide what, how much, and in which manner they share their data, guarantee their privacy and security, and retrieve a fair share of the value their data generates. In such a platform, a personal data cataloguing system is presented to the data owner with its possible associated access policies, to allow the discovery of this kind of data and, at the same time, allow for control that only those who have been specified by the owner have access to them, under the terms previously established. This access control policies system opens the doors to new possibilities in the exploitation of personal data and even proposes a starting point to establish usage control policies to personal data. The KRAKEN project[6] works also on a similar direction offering as one of its main pillars the self-sovereign identity (SSI) solution which essentially is a user-centric access control to data, where the data owner controls their data. The KRAKEN approach is being piloted in two high impact domains, health and education, and contributes to data spaces by providing tools and solutions that can preserve the privacy and assure security, trustability, data integrity, and confidentiality, when data are shared between different stakeholders in a data space, and even between different data spaces.

From an architectural viewpoint, an access control service is composed by an access policies editor, envisioned as part of a data sharing configuration template, as a means for the individuals to define the conditions under which their data will be shared, and an access policies engine, the enabler of the

[6] "KRAKEN - broKeRage And marKEt platform for persoNal data." https://krakenh2020.eu/ (accessed Jul. 22, 2022).

access to the data in case the comparison between the attributes of the data seeker meets the conditions for sharing the data.

6.2.4 Data owners consent management

Consent and adherence to legislation is a very important aspect when it comes to data collection, and contrary to what many think, the technical implementation of such procedures is a very tough task, as it is very hard to offer a highly usable and enjoyable platform to users and, at the same time, tick all the boxes coming from the legal world.

The smashHit[7] project offers a platform that provides functionalities to different actors aiming to get consent creation, management, and observation functionalities. The smashHit platform for this reason includes components[8] such as the following:

- Semantic models of consent and legal rights – This component represents the used ontologies and resulting semantic data models used for the description of legal rights, consents, context and automatic contracting rules, and terms and conditions. The different ontologies are linked by a top-level ontology that describes very basic concepts to be extended by additional linked sub-ontologies.
- Context sensitivity support tool – This component adds context-sensitive features to the overall smashHit framework and enables a context-dependent behaviour of the system.
- Consent certification – This component includes functionalities regarding the life cycle of the consent certifications, i.e., the consent certification creation, consent management, and consent distribution among the parties.
- Data use traceability – This component will allow to find data leakages or misuses by using data fingerprinting or watermark techniques and provides data owners the power to manage their data contracts.
- Contract support tool – This component provides automatic contraction functionalities that enable automatic consent document generation and execution, as well as semantic consent representation and visualisation.

---

[7] https://www.smashhit.eu

[8] https://www.smashhit.eu/wp-content/uploads/2021/03/smashHit_D1.3_Public_Innovation_Concept_v100.pdf

- Consent tracing functionality will be provided, enabling the identification of broken consent chain and guaranteeing that the data exchanged are in agreement with the data consented by data owner.

6.3 Preserving Data Privacy and Data Quality Simultaneously

Though, in the recent years, many methods have surfaced regarding privacy preservation, many of them are a double edge sword when it comes to preserving the quality of the data, as the more privacy preservation measures are applied, the more difficult it becomes to process data and extract highly reliable and accurate analytics results. In this section, we focus on technologies that are already there when it comes to preserve the privacy of individuals and that do have an active research community that pushes forward innovations which aim to disengage data privacy from data quality.

6.3.1 Data anonymisation

Anonymisation is a core enabler for data sharing, especially when one works with personal, sensitive data.[9, 10] The anonymisation process renders data non-identifiable, such that the probability of re-identifying data sharers/individuals in the data is rendered sufficiently low. Identifiability can be viewed along a spectrum. As the data are increasingly transformed, the identifiability of the data is gradually reduced until it reaches a level that is below the applicable anonymisation threshold. At this point, the data are no longer identifiable. Statistical and machine learning mechanisms can be used to anonymise data so that the true values of personally identifiable information (name, age, gender, ethnicity, etc.) are effectively obfuscated and/or transformed (location data). Anonymisation can be incorporated into:

- a "privacy by design" approach to provide improved protection for data sharers;

[9] "Sharing Anonymized and Functionally Effective (SAFE) Data Standard for Safely Sharing Rich Clinical Trial Data." https://www.appliedclinicaltrialsonline.com/view/sharing-anonymized-and-functionally-effective-safe-data-standard-for-safely-sharing-rich-clinical-trial-data (accessed Jul. 25, 2022).

[10] "Guidance Note: Guidance on Anonymisation and Pseudonymisation," *An Coimisiún um Choisant Sonraí. Data Protection Commission*, 2019, Accessed: Jul. 25, 2022. [Online]. Available: https://www.dataprotection.ie/sites/default/files/uploads/2019-06/190614%20Anonymisation%20and%20Pseudonymisation.pdf

- a "data minimisation" strategy aimed at minimising the risks of a data breach.

Both approaches are valid and, in many cases, are combined. Nevertheless, the main challenge in existing anonymisation approaches remains how to manipulate the datasets so that they do not lose the essential information that they hold (from a data scientist perspective), but at the same time satisfy the requirements of the privacy guarantees they are offering to their end-users.

Despite datasets not containing any personally identifying information (PII), such as name, address, etc., individuals can be identified through their quasi-identifiers (QIs). QIs are the attributes whose combination can serve as a unique identifier for individuals. The safe-DEED[11] project offers a demonstrator that is able to perform de-anonymisability analysis of a dataset and check the above-mentioned factors.

6.3.2 Secure data analytic services

Another method to simultaneously support high data privacy guarantees whilst retaining analytics results of high value is the establishment of analytics services that are able to run in secure environments, such as isolated data spaces or on-premise or private hosted cloud infrastructures.

One of the recent infrastructures in this field is the secure analytic host service (SEAS)[12] created in the scope of the DataVaults project to set up and deploy a playground and visualisations host to execute and visualise the different machine learning experiments. The main idea of the SEAS is to allow the user to create a friendly environment which can run in a very user-friendly manner, creating a secure environment where analytics can be executed, as this environment will be ultimately chosen by the user; thus, no third parties are involved.

The SEAS component could be divided into two subcomponents. The first one is the "service analytic host", a component that manages the deployment and set-up of the playground and visualisation host. This is a component that is offered as a central service (in the case of the DataVaults project, this is hosted in the main DataVaults cloud platform). The user/data science/ data seeker defines where he/she wants to deploy the "playground and visualisation host" (see the following subcomponent with regard to choosing

[11] https://www.safe-deed.eu

[12] https://www.datavaults.eu/datavaults-components-for-data-sharing-value-generation-and-intelligence-2-3/

the server, the ports, the datasets, etc.). Technologies such as sbt play[13] as the framework to create the graphical user interface, Java for the setup, and Docker[14] and Ansible[15] to deploy in an automated way the playground and visualisation host is used to enable this component.

The second subcomponent is the "playground and visualisation host". This allows the user to run, track, and visualise the machine learning experiments. This component could be deployed in another server or cloud service outside the operational environment of the "service analytic host", while it could be even deployed in-house, on the user's laptop using a ZIP file. MLflow[16] as a platform for the machine learning lifecycle is the technology used to track and run machine learning experiments. Apache Superset is an exploration and visualisation platform of data that allows the user to understand their data in a better way. Both MLflow and Apache Superset[17] will communicate using a PostgreSQL database.

6.3.3 Data management technologies

Data management in smart cities is a big and complex task. Data gathered and managed by the modern IT services in cities have all characteristics of big data: high velocity, high volume, high value, high variety, and high veracity. In addition to that, the data managed by cities' public administrations are very sensitive and, in many cases, are personal data of the citizens. The modernisation of the IT services in public administrations is a slow, continuous, and never-ending process. There are many good data management systems on the market, but their deployment in the existing complex infrastructures is always a huge challenge. The applications and underlying data management systems of several historic generations often are used in parallel and need to work together. Maintaining the interoperability between them remains a permanent challenge. The best precondition for it is always to follow recognised open standards and ensure the availability of APIs at all services in the public IT infrastructure.

One of the important topics in the context of smart cities is the integration and harmonisation of data coming from different sources. The data create a basis for optimisation of the existing processes and creation of new services, but they can be very heterogeneous. They may have different

[13] https://www.playframework.com/documentation/2.8.x/BuildOverview
[14] https://www.docker.com
[15] https://www.ansible.com
[16] https://mlflow.org
[17] https://superset.apache.org

structure and semantic, which may be not easily understandable by anyone apart from the creator of the service generating the particular data. This is why the documentation of data formats and APIs is very important. The data can be harmonised on the basis of a common data model, which would cover the needs of the planned informational system. In some cases, it is sufficient to harmonise data on the level of data description (metadata) without modifying data itself. Data market place is such an example. The data shared and exchanged through a market place may have different semantic and structure, but the description format of data has to be aligned to make the data discoverable and present the information about data to the data seeker. The best practice in designing data models for metadata is to create them using the semantic web technologies and, in particular, the standardised vocabularies.

Amongst the key technologies in data management is data cleaning, which, amongst others, includes operations such as data validation, data cleaning, and data verification. The aim of such technologies is to offer to all engaged parties the ability to identify all the incomplete, incorrect, inaccurate, or irrelevant parts of this data, and then replace, modify, or delete the dirty or coarse data in order to result in high-quality datasets, which have an improved added value for the data consumers.

One example of such a technology offering comes from the PolicyCLOUD[18] project which delivers a data cleaning component that incorporates algorithms and techniques for detecting and correcting (or removing) corrupt or inaccurate records from diverse types of collected data,[19] including also other intelligent features that have to do with the level of customisation in terms of rules for validation, cleansing, etc., tailored to the needs of each different party that exploits this component.

Another example is that offered by the PIMCity[20] project, which is also offering novel tools for data management.[21] The data knowledge extraction (DKE) component offers the means to extract knowledge from the raw data implementing machine learning and big data solutions. One of the biggest challenges here is the creation of value out of the raw data. When dealing with personal data, this must be coupled with privacy preserving approaches, so that only the necessary data are disclosed, and the data owner keeps the

---

[18] https://policycloud.eu

[19] "Data Cleaning - Policy Cloud." https://policycloud.eu/services/data-cleaning (accessed Jul. 25, 2022).

[20] PIMCity – Building the next generation personal data platforms." https://www.pimcity-h2020.eu/ (accessed Jul. 25, 2022)

[21] https://www.pimcity-h2020.eu/software/

control on them. The DKE consists of machine learning approaches to aggregate data, abstract models to predict future data (e.g., predict user's interests in recommendation systems), and fuse data coming from different sources to derive generic suggestions (e.g., to support decision by users, providing suggestions based on decisions taken by users with similar interests). Another component, the data portability and control (DPC) tool allows individual users to migrate their data to new platforms, in a privacy-preserving fashion. More specifically, it provides methods for extracting data from one PIMS (e.g., bank data through the TrueLayer API), process it by filtering out sensitive information or user-inputted data (e.g., remove login credentials or debit card numbers), and outport it into other modules, a new PIMS (e.g., EasyPIMS[22]) or an exported file in a common data interchange format, e.g., JSON.

PIMCity also offers tools for data provenance and aggregation. *Data provenance module* OpenAPI allows developers to insert watermarks of ownership in the datasets they share in the marketplace. In general, this component is used internally by the PDK and developers that are in need of controlling data ownership even after a dataset has left the platform. This is done by embedding difficult-to-remove watermarks into the datasets. When it comes to aggregation, the data aggregation (DA) tool enables data owners that hold a bulk of their users' data to aggregate and anonymise them. This allows sharing these data in a privacy-preserving way.

The Snap4City platform,[23] which is powering the REPLICATE[24] and is operative with services and data of several cities including Florence, Helsinki, Antwerp Valencia, Venezia, and Roma, includes also a set of tools/services[25] offering a secure and privacy respectful solution for data management, used in several scenarios and functionalities of the platform, working with "my personal data type" according to GDPR, extracting personal data from IoT devices of users, protecting the storage for personal data, etc.

6.3.4 Data models and interoperability

Data management is closely related to data models, as the latter are structures that are in a position to support interoperability in data management systems.

---

[22] https://www.easypims.com

[23] "Snap4City" https://www.snap4city.org (accessed Jul. 25, 2022).

[24] "REPLICATE's Final Results and Impacts – Replicate Project EU." https://replicate-project.eu/the-replicate-projects-final-results/ (accessed Jul. 25, 2022).

[25] "US11. Using tools/services of a secure and privacy respectfully solution - Snap4City." https://www.snap4city.org/drupal/node/166 (accessed Jul. 25, 2022).

When it comes to personal data, the DataVaults project has followed this approach as its data model is defined as a profile of the general data catalogue vocabulary (DCAT[26]). DCAT is a W3C recommendation providing an RDF vocabulary to facilitate interoperability between data catalogues in the web. The DataVaults data model reuses the main parts of the DCAT and extends it with the classes necessary to describe the personal data and domain-specific data of the DataVaults demonstrators. This makes it interoperable with high number of other data models based on the DCAT. The DataVaults data model describes a holistic personal data value chain addressing all the aspects of personal data management. This includes data protection, security, GDPR compliance, IPR management (compensation schemes, etc.), and representation of the main value flows in data marketplaces. It is based on existing semantic web standards that are very important for potential interoperability with other data models and for the model reuse.

A catalogue in DataVaults data model represents the top-level class assigned to an individual. The metadata of a catalogue as defined in the core DCAT vocabulary contains properties that allow description of profiles of individuals. There are two ontologies commonly used for describing an individual's master file data. They have many overlapping properties. In DataVaults, the focus has been laid on the friend of a friend (FOAF) ontology, which is complemented by the vCard ontology.[27] All FOAF-related fields are added via the dct:publisher property and encoded as a foaf:Person element. All vCard-related values can be found as a vcard:Individual attached via the dcat:contactPoint property.

The importance of having proper and interoperable data models is also pinpointed by the i3-MARKET project,[28] as the availability of common data models is a key enabler for establishing a scalable data economy. As such, i3-MARKET offers methods to access in a decentralised manner the semantic descriptions of the offered data assets in order to enable data discovery across today's silos. This enables federation among the individual data spaces and marketplaces, without the need of central control or coordination that has to be trusted by all parties. This is made a reality by using a secure semantic data model repository that enables data consumers to efficiently discover and access data assets (due to precise semantic queries) and integrate the data into their applications/services (based on a common

---

[26] https://dvcs.w3.org/hg/gld/raw-file/default/dcat/index.html

[27] https://www.w3.org/wiki/Good_Ontologies

[28] "i3-MARKETArchitectureandApproach-i3Market." https://www.i3-market.eu/i3-market-architecture/ (accessed Jul. 25, 2022).

understanding of the meaning of the data). In this way, independent data providers and consumers are enabled to exchange and use data in a meaningful way – without prior information exchange. We build on semantic data models defined and make them accessible via a public data model repository.

The PolicyCLOUD's interoperability component aims to enhance interoperability via the utilisation of linked data technologies, such as JSON-LD, and standards-based ontologies and vocabularies, coupled with the use of powerful tasks from the domain of natural language processing (NLP), in order to improve both semantic and syntactic interoperability of data and datasets. Through these coupled technologies and methods, the interoperability component seeks to provide a state-of-the-art approach to achieve interoperability in data-driven policy making domain. SemAI[29] entitles this hybrid approach and is a combination of commonly used semantic techniques coupled with the utilisation of NLP tasks and methods. The main goal of this hybrid mechanism is to design and implement a holistic semantic layer that will address data heterogeneity. SemAI introduces a multi-layer and hybrid mechanism for semantic interoperability across diverse policy-related datasets, which will facilitate semantic interoperability across related datasets both within a single domain and across different policy-making domains. This requirement relates to local–regional public administrations and business domain, but it also goes beyond the national borders as it also seeks to invoke a language-independent hybrid mechanism. To this end, this hybrid approach aims to enhance both semantic and syntactic interoperability of data based on the aggregation, correlation, and transformation of incoming data according to the defined schemas and models. The knowledge that is derived from these processes, shaped in a machine-readable way, can be used later from other tools for providing big data analytics, i.e., sentiment analysis, etc.

6.3.5 Digital twins for privacy preservation

The role of models in a digital twin (in addition to data) is to interpolate spatially where no direct measurement data is available; to integrate the presentation and analysis of complementary datasets; to cross-correlate data from different domains; to infer properties/attributes that are not directly measured;

[29] "SemAI: Enhanced Data Interoperability - Policy Cloud." https://policycloud.eu/services/semai-enhanced-data-interoperability (accessed Jul. 25, 2022).

to convert measurements and state info into KPIs; and finally to extrapolate (predict) business as usual (BAU) and what-if scenarios.[30]

As an example of digital twins for privacy preservation, one could have a look at the DUET project. The project has published the DUET-Cell architecture[31] as a plug-in interface to create a digital twin platform for urban regions where it becomes possible to add new (possibly third-party) data sources to an existing digital twin case, add simulation models to an existing case for comparison or to be used as input for another model adding visualisation clients, and finally extend the digital twin ontology to support more city domains or expand existing ones.

6.3.6 Cryptographic solutions for data privacy

Cryptography is amongst the key enablers to achieve data privacy and is one of the cornerstones when it comes to data protection. For this reason, all existing and emerging platforms do support cryptography out of the box, as one of the many features they employ to protect the data they hold. In the last years, many innovations in the area of cryptography have surfaced and such platforms do consider, or even integrate novel approaches which, at the same time, provide very high trust guarantees and, at the same time, allow various operations over the data, without jeopardising trust or privacy.

As an example of using such state-of-the-art technologies, we suggest KRAKEN[32] that creates a data-analytics-as-a-service component by using secure multi-party computation (SMPC) and functional encryption (FE) for performing computations on protected data. SMPC guarantees confidentiality preventing the data provider, the data processor (the service), and the consumer from learning about the plain data or the sensitive data. Only the consumer is able to learn the statistical analysis result. Additionally, the authenticity and integrity are assured by using zero knowledge proof (ZKP). Sharing data between producers and consumers involves ensuring a secure end-to-end data sharing. Also, ensure that only the consumer has access to the final data or result implies a fine-grained access control. Regarding

---

[30] "DUET D3.4 Smart City domains, models and interaction frameworks v2," *DUET Consortium*, 2021, Accessed: Jul. 25, 2022. [Online]. Available: https://www.digitalurbantwins.com/_files/ugd/68109f_d4c08a03f34e481695977b4fd2577b16.pdf

[31] "Open Technical Approach: T-Cell Architecture for DUET Digital Twins." https://www.digitalurbantwins.com/technical-approach (accessed Jul. 25, 2022).

[32] "Deliverables - Kraken." https://www.krakenh2020.eu/resources/work-package-deliverables (accessed Jul. 25, 2022).

the use of cryptographic methods on self-sovereign identity (SSI) solution, the extension of the SSI solution with ZKP allowing the citizen to disclose part of their attributes in a privacy-preserving way is being investigated in KRAKEN.

Besides ZKP, the use of digital signatures, proxy re-encryption, or attribute-based encryption (ABE) assures a secure end-to-end data sharing and authenticity and confidentiality of data analytics. This approach is being used in KRAKEN as well as in the DataVaults and in the ASSURED[33] project, where in the latter, ABE is handled by hardware TPM devices (of the stakeholders) that have embedded the list of attributes needed for the encryption and the decryption of the data, instead of the attributes being held by a centralised third-party authority.

6.3.7 Artificial intelligence threat reporting and response systems

As we are moving into a world of more and more connected infrastructures, it is essential to also come up with more intelligent ways to detect threats to privacy, and cybersecurity surely plays a very important role in this endeavour.

As an example, one could consider the IRIS[34] project, which addresses the challenge of protecting IoT and AI-driven ICT systems from cyberthreats and cyberattacks on their operation and privacy. The IRIS concept is proposed as a federated threat intelligence architecture that instates three core technological and human-centric components into the threat intelligence ecosystem. First, there is the collaborative threat intelligence that forms the nexus of the IRIS framework and core component of the architecture enhancing the capabilities of the existing MeliCERTes platform by introducing analytics orchestration, an open threat intelligence interface and an intuitive threat intelligence companion. All this supported by a data protection and accountability module. Second, there is the automated threat analytics that collects and supplies key threat and vulnerability assessment telemetry and responds to received intelligence, initiating autonomous response and self-recovery procedures. The third component is the cloud-based virtual cyber range, which delivers an immersive virtual environment for collaborative training exercises based on real-world environment platforms (and digital

[33] https://assured-project.eu

[34] "IRIS H2020 project." https://www.iris-h2020.eu/ (accessed Jul. 22, 2022).NOTE-Not to be confused with the IRIS project referred to in Chapter 2 which is https://irissmartcities.eu/

twin honeypots), providing representative adversarial IoT and AI threat intelligence scenarios and hands-on training.

6.4 Information Delivery on Privacy Metrics and Data Content and Value

Aside from the tools and methods used to allow engaged stakeholders to secure their data and build the necessary barriers to safeguard privacy and enable trust, there is also the need to explain to data owners how their data and, at the very end, their privacy is protected, what they can find out from their own data themselves, as well as how much value their data hold. In Section 6.4, we explore some of these technologies.

6.4.1 Privacy metrics and risk management and privacy metrics for personal data

Risk management contains both the notion of monitoring as well as evaluating the risks related mainly to privacy. The main root of security risks is the sharing of a dataset that contains private information, and a risk management system can offer two different ways to analyse this risk.

First, it is in the position to provide an overview of privacy risks and the privacy exposure to the platform administrator on the basis of the datasets that have been shared to the platform. This is based on a model used to serialise all information that can describe the datasets and then perform the calculation of relevant risks. The above privacy assessment is based on the analysis of asset chains in order to estimate the impact from the interconnections of the assets and datasets and to provide a single estimation for the overall impact of a specific asset/vulnerability combination. Risk management also provides the ability to identify and add new known vulnerabilities to enrich the risk knowledge base of the platform.

Second, it can provide to the end-user the calculated risk regarding the datasets that the end-user has added to the platform, and it also warns her/him of the risk exposure of a dataset that she/he is trying to share, in order to enable more informed decisions regarding the amount of personal information the user is comfortable sharing.

The DataVaults project is delivering such a risk management component where input is taken from both the datasets that an owner want to share as well as from the dataset that has been already shared by the same individual, to assess a unified privacy risk exposure metric.

The PIMCity personal information management systems (PIMS) development kit[35] (PDK) also similar components are provided, such as the following:

- Personal data safe (P-DS) is the means to store personal data in a controlled form. It implements a secure repository for the user's personal information like navigation history, contacts, preferences, personal information, etc.
- Personal privacy metrics (P-PM) represent the means to increase the user's awareness. This component collects, computes, and shares easy-to-understand metrics to allow users to know how a service stores and manages the data.
- Personal consent manager (P-CM) is the means to define all the user's preferences when dealing with personal data. It defines which data a service is allowed to collect, process, or can be shared with third parties.

6.4.2 Personal data analytics

Analytics at the side of an individual/citizen can offer to that person two main things in a platform that handles and manages personal data. On the one hand, they can allow her/him to have a quick view of the data she/he generates using cumulative and predefined graphs and charts. Doing that, a citizen is able to better understand the data she/he generates and can have some high-level insights of what other stakeholders might discover if she/he decides to share her/his data. Think, for example, that instead of seeing just bio signals, one could see the daily or hourly heart rate, or instead of having a long list of things she/he searched for in Google, to have her/his queries categorised based on identifying how much this person is, for example, querying for sport facilities, real-estate agents, and shopping.

On the other hand, the ability to export analytics allows the individual also to engage in data sharing activities by just sharing these analyses, which increase the privacy level of the data subjects, as instead of sending out raw data, already pre-processed data are provided.

The PIMCity personal information management systems (PIMS) development kit (PDK) includes a personal privacy preserving analytics (P-PPA)

---

[35] https://www.pimcity-h2020.eu/about-the-technology/pimcity-development-kit/

which allows extracting useful information from data while preserving users' privacy and leverages concepts like k-anonymity and differential privacy.

The DataVaults project showcases how all of the above can be delivered to an individual by having a lightweight analytics engine running at the side of the user, which will rely on predefined algorithms and data categories linked to the data sources that a citizen is willing to share with other stakeholders.

6.4.3 Data valuation

Understanding the real value that is included within a dataset is a crucial point for all engaged stakeholders in the value chain of the data economy.

The Safe-DEED project provides tools to facilitate the assessment of data value, thus incentivising data owners to make use of the cryptographic protocols to create value for their companies and their clients.[36] The data valuation applications, as part of the Safe-DEED demonstrator, describe the initial implementation of the data valuation component (DVC). The supported algorithms are selected regression, classification, and clustering algorithms (at ADAS level), and a rule-based algorithm for generating the economic value of the input data set (at S2VM level).

The PIMCity PIMS development kit (PDK) also includes data valuation tools. These tools, from the market perspective (DVTMP) module, leverage some of the most popular existing online advertising platforms to estimate the value of the audience, while from the user perspective (DVTUP), it provides estimated valuations of end-users' data for the bulk dataset they are selling through the marketplace.

6.5 Data Platforms

All of the above-mentioned technologies are being integrated into data platforms, either industrial ones or personal data platforms, which aim to bring different stakeholders together to make the notion of the data economy a reality, respecting both data owners, data consumers, and all other engaged stakeholders such as data brokers, etc.

In this subsection, some key technologies for the realisation of such platforms are presented as well as some examples of such marketplaces.

[36] "Safe-DEED Valuation tool." https://demo.safe-deed.eu/ (accessed Jul. 25, 2022).

6.5.1 Secure and trusted data communication channels

One of the main features that data platforms should include is that of secure data exchange which is a key factor in generating the trust between the transaction parties.

IDS is one of the most prominent initiatives for creating connectors for secure data sharing. The project DataPorts[37] which works on secure managing and sharing transportation and logistics data among trusted stakeholders and using these data to offer novel AI and cognitive tools to the seaport community, has based its architecture on the IDS concept. In more detail, the DataPorts platform is based on the concepts from the IDS reference architecture designed for secure data exchange and trusted data sharing. These concepts are represented by components in the DataPorts platform. The IDS reference architecture addresses such strategic requirements as enabling: trust, security and data sovereignty, ecosystem of data, standardised interoperability, data market, re-use of existing technologies, and contribution to standardisation.

To enable secure end-to-end data exchange, the i3-MARKET project makes use of secure and trusted APIs to allow data spaces and marketplace providers to obtain identities, to register data assets, to fetch their semantic descriptions, to create and sign smart contracts, to make payments, etc. This ensures complete openness, i.e., that any data space or marketplace provider can connect its local ecosystem with the global i3-MARKET data market ecosystem.

Secure communication is also established by peer-to-peer connections, as showcased by the InteropEHRate[38] project, where key health data is managed in "patients' hands", i.e., through smart EHRs (S-EHR) on mobile devices. In InteropEHRate, data is always transferred via highly secure channels including a direct device to device (D2D) communication. It is developing open interchange protocols supporting patient-centred exchange of health records between patients, healthcare actors, and researchers. It is guided by a future-proof prospective where most citizens will own mobile health repositories, called smart EHRs (S-EHRs), managing a wide range of their own personal health data on smart devices, regardless of whether the data have been produced by health professionals, any sensor, device, or the citizens themselves. As such, InteropEHRate gives citizens and patients a meaningful choice for the method their health data is being stored and exchanged, such that they can benefit from its usage everywhere it is needed and gain also more control making the process fully compliant with the GDPR. This choice

[37] https://www.dataports-project.eu/

[38] https://www.interopehrate.eu/about/

is made possible by developing an innovative health data storage on smart phones. Citizens can also initiate themselves standardised communications and transfer data, e.g., to hospital outpatient services by using secure remote connections via the internet, and, in addition, by building new connections without the usage of internet between the citizen's device and the healthcare provider systems using short-range device to device communication.

6.5.2 Immutable ledgers and smart contracts

Data exchange in novel marketplaces takes advantage of the recent innovations in the technological domain of blockchain, and, as a result, it is nowadays a commodity to store data sharing transactions in blockchain ledgers and employ smart contracts that can automate many of the necessary activities that are included in data exchange.

The i3-MARKET project makes use of immutable and auditable smart contracts for the trading of data assets across data space and marketplace boundaries. All stakeholders, namely data providers (for confirmation of the offer and its conditions, e.g., license, price, and SLAs), data consumers (for agreement of the contract conditions), and data owners (for consent to the data exchange) must sign these contracts. This developed solution can also be adopted by individual marketplaces for handling local contracts. Similar solutions are also followed by other projects, such as DataVaults where a double ledger approach is provided in order to keep data owners and data consumers apart, with the platform playing the role of the data broker, or as in BEYOND[39] project, where a similar ledger approach is used which also enables multi-party contracts.

KRAKEN[40] developed a decentralised solution such as SSI which provides a decentralised user-centric approach on personal data sharing; it uses verifiable credentials (VCs) with different levels of assurances for accessing online services. KRAKEN also provides a trusted environment comprising different trusted registries, in order to be ready for being aligned with EBSI[41]/ESSIF[42] and leveraging their capacities, supporting trustability in SSI solution. Also, KRAKEN provides a public DID infrastructure supporting access to different data ledger technologies (DLTs) and the management of public

[39] https://beyond-h2020.eu

[40] "Deliverables-Kraken."https://www.krakenh2020.eu/resources/work-package-deliverables (accessed Jul. 25, 2022).

[41] https://ec.europa.eu/digital-building-blocks/wikis/display/EBSI/Home

[42] https://www.eesc.europa.eu/sites/default/files/files/1._panel_-_daniel_du_seuil.pdf

DID, providing the associated services for creating/generating, publishing, and then using the public DIDs. The ledger uSelf component developed in the context of KRAKEN comprises two subcomponents. One is the SSI mobile app for managing the VC and key material, allowing the citizens to use their own smart devices for identifying themselves using the SSI solution. The mobile app provides a friendly graphical user interface designed to adopt and simplify the SSI workflows. The other is the ledger which is used as is used as a broker for integrating service providers (SPs), reducing the complexity of the protocols and processes associated with the SSI solution by providing a simplified interface. With the SSI app and the broker, the SP can use SSI key material for onboarding and login processes. A backup/synchronisation system is integrated with the SSI solution allowing the citizen to use several devices with the same credentials and recover their credentials in case the smart device is lost or stolen.

6.5.3 Crypto wallets

Crypto wallets are used to allow entities (be it individuals or organisations) to have a place where they can digitally retain information relevant to their crypto currency. As crypto currency is, of course virtual, these wallets are used to store cryptographic keys to the crypto currency that each user has. This means that the crypto currency is not actually stored in the wallet but leaves in the blockchain, and the role of the wallet is to hold securely the private cryptographic keys of everyone to be able to access their crypto currency. It is therefore of outmost importance that users do not lose access to their crypto wallets, as in such a case, there is no way to be able to regain control of the crypto currency the user has linked to that specific wallet.

Crypto wallets are highly important in online trading and can be used in operations such as data sharing. They offer an initial degree of privacy, as a user is identified by a hash; however, it is possible by exploring the transactions of a party in a blockchain network to reveal information about his identity. There are also approaches proposed, such as the one of the DataVaults project, which suggest using intermediate wallets owned by trusted third parties, to play the role of the broker between two or more entities, maximising the privacy guarantees and anonymity of the transaction parties, as they later will be actually transacting only with the broker who is the only one knowing their true identities of the different parties.

Transactions in the i3-MARKET project are based on a crypto currency/token as a means to provide a transparent, cost-efficient, and fast payment solution for trading data assets among the participating data spaces and marketplaces. The crypto token is used to incentivise data providers to offer their data assets and thus accelerate the European data economy. The solution is designed in a way that the participating data spaces and marketplaces can also use the tokens as internal payment medium.

6.6 Other Supporting Initiatives

In addition to the outputs from the projects that have been referred to above, there are a variety of other initiatives offering supporting frameworks. Many fit into both camps, providing valuable technologies as well as providing a more general supporting function, such as RUGGEDISED and SmartEnCity.

6.6.1 EUHUBS4DATA

The federation aims at creating a solid ecosystem, bringing together relevant European initiatives around the data economy, fostering collaboration among those initiatives towards the objective of common European data spaces, attracting SMEs and start-ups to use and benefit from the federated services and data sources, and raising awareness in society about the benefits of data-driven innovation.

In fulfilling this aim, a catalogue of services was offered, datasets were made available, and courses on the offer have been provided.

The "catalogue" of EUHUBS4DATA[43] comprises:

- 172 services offered by the membership, ranging from "access and support to big data-AI stack environment" through to "very large speech dataset to build up automatic speech recognition and text to speech deep models".
- 160 datasets made available by the membership ranging from "air quality" through to "world bank open data".
- 80 courses offered ranging from "add-ons to courses: workshop on ML for energy, manufacturing, fintech, transportation, healthcare, etc." through to "strategies for data-based business models and hands-on development".

[43] "EUHubs4Data - European Federation of Data Driven Innovation Hubs." https://euhubs4data.eu/ (accessed Jul. 25, 2022).

6.6.2 MyData

The MyData Mission Statement[44] is: "We help people and organisations to benefit from personal data in a human-centric way. To create a fair, sustainable, and prosperous digital society for all".

"MyData Global aims to empower individuals by improving their right to self-determination regarding their personal data. The human-centric paradigm strives for a fair, sustainable, and prosperous digital society, where the sharing of personal data is based on trust and a balanced and fair relationship between individuals and organisations".

6.6.3 Solid Flanders

"Solid[45] is a specification that lets people store their data securely in decentralised data stores called Pods. Pods are like secure personal web servers for your data.

Any kind of information can be stored in a Solid Pod. You control access to the data in your Pod. You decide what data to share and with whom (be it individuals, organisations, and/or applications). Furthermore, you can revoke access at any time. To store and access data in your Pod, applications use standard, open, and interoperable data formats and protocols".

6.6.4 Big value data association (BDVA)

"The Big Data Value Association –BDVA, (from 2021, DAIRO - Data, AI and Robotics), is an industry-driven international not–for-profit organisation with more than 230 members all over Europe and a well-balanced composition of large, small, and medium-sized industries as well as research and user organizations. BDVA/DAIRO focuses on enabling the digital transformation of the economy and society through Data and Artificial Intelligence by advancing in areas such as big data and AI technologies and services, data platforms and data spaces, Industrial AI, data-driven value creation, standardisation, and skills. The mission of the BDVA is to develop the Innovation Ecosystem that will enable the data and AI-driven digital transformation in Europe delivering maximum economic and societal benefit, and, achieving and sustaining Europe's leadership on Big Data Value creation and Artificial Intelligence".[46]

---

[44] https://www.mydata.org/

[45] https://solidproject.org/about

[46] https://www.bdva.eu/about

6.7 Looking into the Future

As the current IT landscape is witnessing dramatic changes which are highly influencing and are also influenced by the changes in our societies, numerous changes are expected to happen in the area of data sharing, privacy, trust, and security. In order for someone to be able to remain at the forefronts of the technology, much effort has to be invested in operations such as technology watch and applied developments and testing of current solutions as those mentioned briefly in the previous subsections.

Regarding this chapter, it is mentioned that the results from all the projects referred in the previous subsections are and will be published on their respective web sites as they appear and will generally include all the supporting tools and manuals for the implementation of these results. The DataVaults "manual", the RUGGEDISED D6.6, the guidance within IRIS, etc., all add to the repository of knowledge for moving forward. In addition, a new platform has been set up by the Commission to also store relevant information.[47]

Chapter 7 will start to address some of the interoperability issues faced when augmenting existing platforms and approaches already taken by a smart city.

Further, Chapter 17 points to the technologies that will become available in the near future, from projects about to start, as given the time scale set by many cities of becoming climate neutral by 2030, there cannot be any unnecessary delay in making use of these emerging technologies if these goals are to be achieved. Advance awareness of what is coming next will shorten the process of adoption.

[47] https://ec.europa.eu/info/funding-tenders/opportunities/portal/screen/opportunities/horizon-results-platform

7

Interoperability and the Minimal Interoperability Mechanisms

Michael Mulquin

MIMs Ambassador, Open & Agile Smart Cities
Email: michael@oascities.org

Abstract

The chapter sets the context of local data sharing ecosystems, where data from many different agencies can be brought together to enable the city to be managed in a more holistic way. It points out that this requires technical, information, and organisational interoperability and provides a list of some of the specific areas where interoperability is needed in such an ecosystem. It then places this within the European Policy Context.

The concept of minimal interoperability is then dealt with as a way of enabling small- and medium-sized cities and communities to put in place "good enough" interoperability mechanisms to enable effective data sharing without requiring excessive time or resources to implement. The chapter closes by reviewing the minimal interoperability mechanisms being developed by Open & Agile Smart Cities that are incorporated within the Living-in.EU initiative.

7.1 The Context – The Local Data Sharing Ecosystem

The people and businesses located in the city are served by a variety of different systems including electricity, water, gas and other utilities, communications, transport and logistics, education, health, and shopping and commerce. All these systems interact with each other, all of them generate data, and all of them require good data to work well.

DOI: 10.1201/9781003399384-8

We can define a smart city as one where increasing amounts of useful data about the city are collected and used by the public administration, by business, and by the citizen, to help the city work better.

A local data sharing ecosystem is a way of describing what is needed to allow the data being gathered by many different agencies within the city to be brought together in a carefully managed way, to provide comprehensive understanding of what is going on in the city, to allow much better city management, and to support citizens in managing their lives more effectively.

This requires interoperability – ensuring that all organisations and systems providing the data follow common standards and protocols. Having many cities around the world following those same interoperable standards and protocols will facilitate the development of a global market in the products and services that utilise and exploit city data.

7.2 Interoperability

ITU defines interoperability as: the ability of two or more systems or applications to exchange information and to mutually use the information that has been exchanged.[1]

Expanding on this, the GridWise Architecture Council[2] considers that interoperability incorporates the following four characteristics:[3]

1. exchange of meaningful, actionable information between two or more systems across organisational boundaries;
2. a shared understanding of the exchanged information;
3. an agreed expectation for the response to the information exchange;
4. a requisite quality of service: reliability, fidelity, and security.

Interoperability is not just about solving different categories of technical issues. For the information to be exchanged and used, the systems involved need to use consistent mechanisms across a number of informational and organisational categories.

---

[1] ITU-T Recommendation Y.101 Global Information Infrastructure terminology: Terms and definitions

[2] The GridWise® Architecture Council (GWAC) was formed by the U.S. Department of Energy to promote and enable interoperability among the many entities that interact with the nation's electric power system.

[3] See: https://gridwiseac.org/pdfs/interopframework_v1_1.pdf

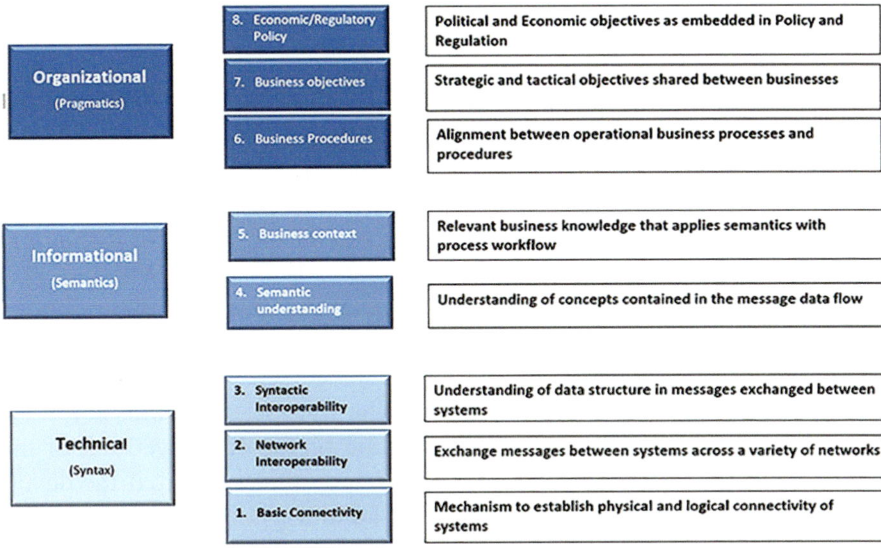

Figure 7.1 Interoperability framework categories.

These different categories can be illustrated by Figure 7.1 which is based on one developed by the GridWise Architecture Council.

As an example, for two organisations to be able to share data relating to individual persons, they both need to comply with a common legal and regulatory framework, such as the European GDPR. In many cases, ensuring interoperability within the organisational and informational categories is more difficult that within the technical ones.

Within local data sharing ecosystems, some of the key organisational and informational issues that need to be handled in a consistent and interoperable way, by all the different data sharing organisations, include:

- knowledge and context information exchange;
- use of consistent data models;
- rules of access and use for data and services;
- protection of rights (personal data, privacy, dignity, equality, etc.);
- transparency in automated decision-making (societal governance of all technology use and deployment);
- security (systems and society);
- management of location data;

- common societal objectives with measurable outcomes towards those objectives;
- interoperability of complex data models, allowing more efficient analytics and impactful exchange of expertise;
- the use of common resource management frameworks.

We will pick up some of these in more detail in the section on minimal interoperability mechanisms.

7.3 The European Policy Context

Interoperable Europe is a new initiative of the European Commission for a reinforced interoperability policy in the public sector. It evolved out of the ISA² funding programme of the European Union that supported the development of digital solutions to enable public administrations, businesses, and citizens in Europe to benefit from interoperable cross-border and cross-sector public services. That programme finished at the end of December 2020.

The issue, of course, is still not solved, and so the European Commission and its partners in public administrations across Europe are now working under the label of Interoperable Europe to continue to enhance interoperability to unlock the potential of data use and reuse for improved public services.

A recent study by the European Commission's Joint Research Centre (JRC) states that improved interoperability could lead to a reduction in the time citizens spend every year with the administration by 25%. This results in time savings of 24 million hours (about 2738 years) and monetary savings in the order of EUR 543 million per year. For business, the savings could reach up to EUR 568 billion annually.

Interoperable Europe will lead the process of achieving these goals and creating a reinforced interoperability policy that will work for everyone. It is committed to introducing a new cooperative Interoperability Policy Directive for Europe that will transform the public administrations and help them in their digital transformation. The initiative is supported by the Digital Europe Programme.

As part of this wider push for interoperability, the Proposal for a European Interoperability Framework for Smart Cities and Communities (EIF4SCC)[4] was published in May 2021. The aim is to focus on the specific

---

[4] https://op.europa.eu/en/publication-detail/-/publication/f69284c4-eacb-11eb-93a8-01aa75ed71a1/language-en

needs and opportunities that interoperability provides in the local context. The proposal is being discussed through the Living-in.EU community and other forums, with a view to its adoption as an official Commission document, based on users' and stakeholders' feedback.

The proposal provides 30 recommendations, along with many helpful case studies, relevant to the development of interoperable local data sharing ecosystems. It is worth highlighting some of these recommendations here.

Recommendation 12 is to "Set-up or consolidate interoperable local data ecosystem(s) that integrate and reuse data in cities and communities by stakeholders, and promotes open standards and open technical specifications, APIs and data models to provide a holistic view of the information. This aims to support the decision-making process and to foster innovation and citizen engagement".

Recommendation 14 is to "Reuse and share solutions, data, tools and services by cooperating with different stakeholders in the design, development, implementation and monitoring phases of service provision at local, regional, national and European levels".

The reasoning behind Recommendation 14 is the need to acknowledge the role of non-public administration actors in service provision in the context of Smart Cities and Communities. As a substantial amount of data and information being generated in the city is out of the hands of public administrations, the governance of any data sharing ecosystem must take a broader view and look beyond the public administration itself.

Recommendation 24 is to "Create more horizontal services towards local data ecosystems, to overcome silos within different domains, by encouraging collaboration and engagement among inhabitants, business, visitors, organisations and city/community administrators".

This points to the importance of making sure that the local data sharing ecosystem is used to support a much more holistic way of managing city services. It needs to move from simply being about using data sharing to improve the way the individual silos are managed, to one where the data is used to highlight new ways of delivering those services in ways that are much more focused around the needs of the citizen rather than on the structures and priorities of the individual departments and municipally owned companies. This points to the key role of the organisational areas of interoperability.

The European Commission recognises that while urban and rural communities in Europe are advancing in their digital transformations, they are not all at the same level. Some have already moved towards an integrated cross-sector approach to exploit the strengths of advanced digital technologies such as digital twins, local data ecosystems, AI, advanced data analytics, high performance computing, and cloud computing. Others have started to invest in their digital transformation but need to accelerate. A third group of communities are at an early stage or have not started at all.

The Digital Agenda addresses all three groups, reinforcing the European capacity for the deployment and scale-up of AI-powered digital twins and enabling local data ecosystems in a large number of European cities and communities including the EU outermost regions and other economically disadvantaged regions.

To make this happen, the Commission has set up the Living-in.EU initiative. The signatories of the "Living-in.eu Declaration"[5] have set ambitious goals to accelerate their digital and green transition. They have agreed on a set of specifications and requirements that should enable interoperability and form the basis of their local data ecosystems and (in the next phase) local digital twins. These are the minimal interoperability mechanisms or MIMs.

7.4 Minimal Interoperability Mechanisms

As we have seen, interoperability enabled by common standards and requirements is a key part of enabling a local data sharing ecosystem.

Minimal interoperability is defined by the ITU as:[6] "The minimal sufficient degree needed to meet a certain requirement for data sharing, use and reuse. NOTE – This is an approach to build a set of modular mechanisms, including information models, across multiple domains, locations and events".

Minimal interoperability mechanisms are the minimal but sufficient capabilities needed to achieve interoperability of data, systems, and services between buyers, suppliers, and regulators across governance levels around the world. By basing the mechanisms on an inclusive list of baselines and references, they can take account of the different backgrounds of cities and communities and allow cities to achieve interoperability based on a minimal common ground.

Implementation can be different as long as crucial interoperability points in any given technical architecture use the same interoperability mechanisms. Each MIM can further define a hierarchy of levels of interoperability based on sectorial needs or the need for tighter integration. The MIMs are vendor neutral and technology agnostic, enabling anybody to use them and integrate

[5] https://www.living-in.eu/

[6] FG-DPM Technical Specification D0.1 Data Processing and Management for IoT and Smart Cities and Communities: Vocabulary [b-FG-DPM TS D0.1]

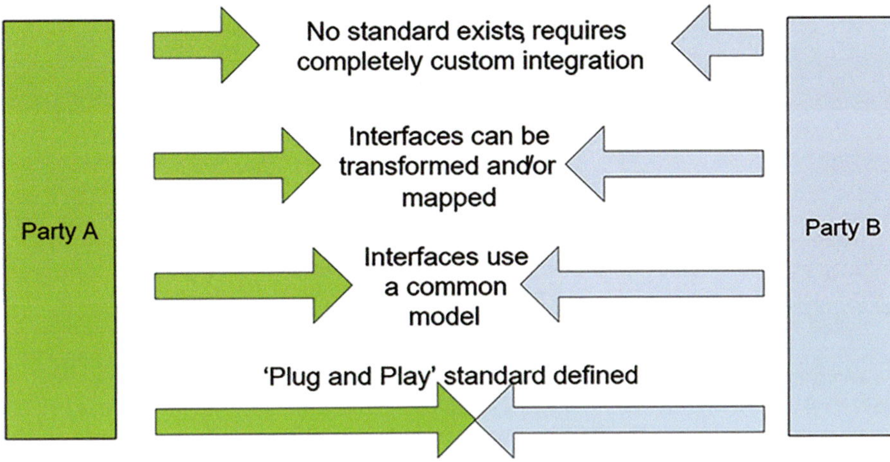

Figure 7.2 Distance to integrate.

them in and between existing systems and offerings, complementing existing standards and technologies.

Minimal interoperability mechanisms reflect the fact that different levels of standards can enable different levels of interoperability, as illustrated in Figure 7.2. Full compliance with detailed and well-designed standards can enable "plug and play" where the different components can be automatically configured to work together merely by "plugging them in" to each other. However, standards-based mechanisms that are not quite so rigorous can still play a major role in supporting interoperability as the following diagram from the GridWise Interoperability Framework illustrates.

The reason why MIMs are necessary is that there are many guidelines and frameworks covering different areas of concern that need to be put in place to enable a fully functional data-sharing ecosystem for smart cities and communities. While this can be managed effectively by larger and well-resourced cities and communities, most small- and medium-sized cities find the complete implementation of all the standards and frameworks a complicated and daunting task.

The MIMs are minimal to ensure no unnecessary complexity or time-to-implement, with the aim that the cost to implement (staff time, software, and hardware) will be affordable by small- and medium-sized cities and cities with limited resources.

MIMs are simple and transparent mechanisms, ready to use in any smart city or community, regardless of size or capacity, even to the national level, global regional level, or globally. The interoperability points assure

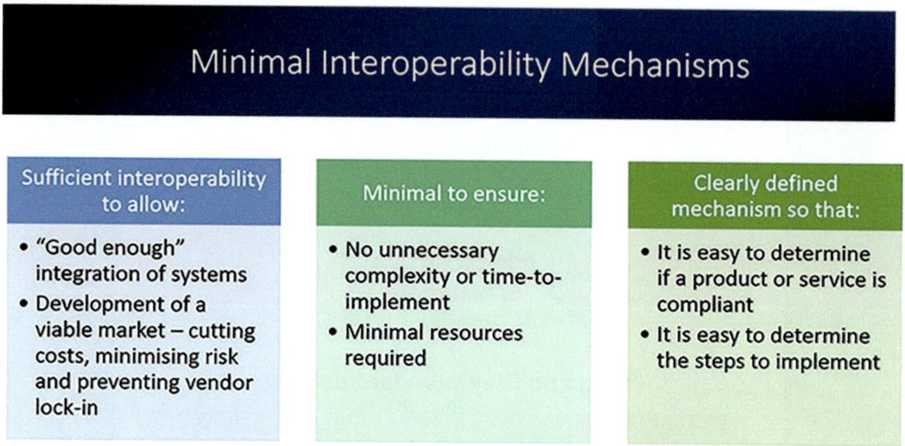

Figure 7.3 Minimum interoperability mechanisms.

the replicability of the solutions built on top of an open platform, as these are decoupled from the specific technological implementations and deployment of the architectural components. Figure 7.3 summarises these minimum interoperability mechanisms.

Essentially, MIMs provide simple, straightforward ways for a city to implement the essential aspects of what is needed to support interoperability within a local data sharing ecosystem. The aim of defining these is to enable the digital capabilities of any city or community to be based on a firm, controllable foundation. By embedding them within a city data platform and data architecture, all stakeholders can be sure that they have the data management and processing capabilities and the interoperability needed.

There are three different types of MIMs, each of which focuses on delivering the minimal but sufficient level of interoperability needed to enable an effective data sharing ecosystem.

1. Where there are existing authoritative standards, MIMs point to their core requirements to enable cities and communities to see immediate benefit in developing the local data ecosystem.

2. Where there are several standards that cover the same ground, the aim will be to identify the lowest common denominator (or the Pivotal Point of Interoperability) that will make it easy to link products and services that comply with those different sets of standards. Pivotal Points of Interoperability is a concept defined by the US National

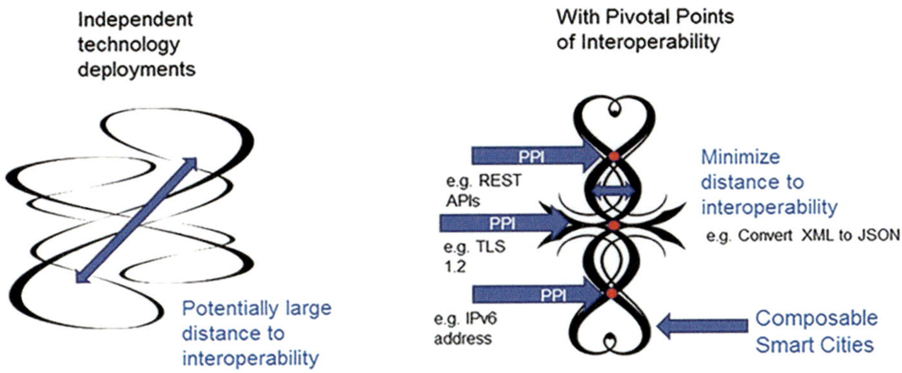

Figure 7.4 NIST pivotal points of interoperability.

Institute of Standards and Technology and used as the basis for the internationally developed IOT Enabled Smart City Framework[7] that NIST facilitated, as illustrated in Figure 7.4.

3. Where there are no existing standards, then MIMs can be developed in partnership with cities and city stakeholders that can act as minimum viable (standards) products. These can then be used as the basis for developing more detailed and comprehensive standards by Standards Development Organisations.

The idea of MIMs has gained widespread support and is included as recommendations in national guidelines, including the Danish Guide to Sustainable Digital Transformation.[8]

ITU Study Group 20 has started the process to define and describe the minimal interoperability mechanism (MIM) approach to developing requirements related to interoperability in local data ecosystems in smart and sustainable cities and communities so that they will be established formally within international standardisation. The work is to:

- provide a definition and common format for the MIMs to provide them with an established role in the standards world;

[7] NIST IES city framework https://s3.amazonaws.com/nist-sgcps/smartcityframework/files/ies-city_framework/IES-CityFramework_Version_1_0_20180930.pdf

[8] Danish Guide to Sustainable Digital Transformation – DS/INF 176:2021 (2021), Danish Standards. Available online: https://webshop.ds.dk/en/standard/M346207/ds-inf-176-2021

- provide a framework to enable MIMs to be developed in a consistent way;
- review the scope of the set of MIMs needed to enable the development of a local data sharing ecosystem, identify any others needed, and provide a process to agree and scope out any further MIMs that may become needed in the future.

7.5 The Individual MIMs

Open & Agile Smart Cities – a network of 168 cities in over 30 countries – is developing a set of 10 MIMs with the aim of covering the full set of requirements to put in place an effective local data sharing ecosystem.[9] Table 7.1 lists them.

Of course, this list may be added to later to cover any gaps that are identified. The process of making sure the full list is covered is underway. That is important as all the MIMs have dependencies on some of the other ones, and having the full list will enable those links to be put in place.

MIMs 1, 2, and 3 are already at a good level of maturity and are being widely specified by cities in procurements, while MIMs 4, 5, and 7 are under development and plans are in place to develop the remaining MIMs.

7.5.1 MIM1 context information management

IoT data provides near real-time information about what is happening in the city and the opportunity to analyse historic IoT data to detect patterns and help identify the causes of problems and how best to tackle them. However, IoT data on its own is not enough.

Air quality data from IoT sensors is itself just a stream of numbers. To make sense of that data, we need to know, for instance, that one sensor is near a busy road, another is in a quiet residential neighbourhood, and another is near a factory. We also need to know what levels of pollution are dangerous, so that warnings can automatically be triggered.

Similarly, information about timings of the data from the sensors needs to be linked with, for instance, information about the weather at that time, or the season of the year, or any event that happened at a specific time that might have had an impact on the readings of the sensors.

[9] See https://mims.oascities.org

Table 7.1 MIMs 1-10.

MIM	Name	Function
MIM1	Context Information Management	Manages the context information coming from Internet of Things (IoT) devices and other public and private data sources, providing cross cutting context data and access through a uniform interface. It therefore ensures comprehensive and integrated access, use, sharing, and management of data across different solutions and purposes. The ETSI standard NGSI-LD is the recommended specification to achieve this.
MIM2	Shared Data Models	Ensures that the datasets in a local data ecosystem use precisely the same machine-readable definitions for key terms to enable datasets to be linked with other sets that add important context information.
MIM3	Ecosystem Transactions Management	Provides functionalities to enable effective matchmaking of urban IoT data sources from providers with respective data consumers, to facilitate trusted exploitation of such data based on enforceable data usage agreements and to secure value flow between these stakeholders.
MIM4	Personal Data Management	Will provide guidance regarding what solutions cities can use to allow citizens/users to control which datasets/attributes they want to share with solution, application, or service providers under transparent circumstances, enabling trust between the different parties.
MIM5	Fair Artificial Intelligence	Will provide tools to cities so that they can have confidence that the AI and the models they use, as well as the goals the AI is programmed to achieve, are fair and transparent, and that they use data from the local digital ecosystem in a fair and transparent way.
MIM6	Security Management	For data to be used in the data ecosystem, it may often need to go through a complex path between where it is generated and where it is finally used. At every stage in that process, it is vulnerable to attack and proper systems need to be put in place to address this. MIM6 will provide cities with the tool to do this.
MIM7	Geospatial Information Management	Will enable cities needed to use sophisticated geospatial data, for instance in 3D modelling and in locating spaces within buildings. In this way, it will enhance the geospatial data handling abilities covered by MIMs 1 and 2.
MIM8	Ecosystem Indicator Management	Will develop consistent measures of the ability of different cities to provide a healthy and effective ecosystem that nourishes digital transformation and supports interoperability of data, systems, and services. This will allow cities to assess their performance against these measures. It will also support the benchmarking of results and practices among comparable peer cities.
MIM9	Data Analytics Management	Will aim to make complex data models interoperable, allowing more efficient analytics and impactful exchange of expertise, to allow cities to leverage each other's successes in data analytics.
MIM10	Resource Impact Assessment	Will aim to develop interoperable capabilities for management and assessment of scarcity and resources related to people, nature, and investment.

So, other data is needed to make IoT data useful. We need context data to make sense of the IoT data to support real-time management of the city – to understand what actions to take as a result of the information from the IoT sensors.

Context data also enables the city to analyse historic data – to understand causes of variations and what issues need to be tackled in longer term in order to deal with any problems. It also allows the city to review potential options for how to tackle those issues so that it can know what different solutions are practical and implementable.

MIM1 points to a tried and tested way to manage context information using the ETSI NGSI-LD standard. This requires context information to be structured around entities that have properties and relationships to other entities. To do this, it requires data models to be described using the Resource Description Framework (RDF) methodology, Resource Description Framework Schema (RDFS), and Web Ontology Language (OWL).

NGSI-LD then describes the APIs needed to link the context data appropriately with the relevant datasets.

7.5.2 MIM2 shared data models

To be able to link the context information to the correct part of any dataset, it is important that the datasets concerned use precisely the same definitions for key terms. For instance, if the one dataset defines "children" as people aged between 5 and 15 and the other dataset defines children as people between the ages of 2 and 12, then a great deal of inaccuracy would result by combining them.

More fundamentally, to enable datasets to be combined automatically, the terms used in each dataset need to be defined in machine readable terms so that the APIs can "understand" how to handle them. Data models are machine readable definitions of key terms.

Finally, the data models need to be in a format that allows the context management API to enable apps to link the context data to the appropriate cells in the original dataset – in other words, they need to comply with the requirements set by NGSI-LD.

7.5.3 MIM3 finding and using the data

MIM3 is the management layer that allows stakeholders:

- to provide data along with relevant information about its content and quality and any terms and conditions for use;

- to provide data processing services along with relevant information and terms and conditions for using the services;
- to find and access the data and data processing services and other services they need and to be able to gain relevant insights into what those data streams/data processing services/data applications consist of and how valuable they can be.

There are various ways to realise this management layer. A standardised way of doing so is provided by TM Forum and Fiware that have created an API suite of specifications for digital marketplaces, named the Business API Ecosystem.[10]

7.5.4 MIM4 personal data management

Personal data management means providing clear and easy usable means for citizens/users to control which datasets/attributes they want to share with solution, application, or service providers under transparent circumstances, enabling trust between the different parties.

Citizens should be able to identify themselves with an ID of their choosing and be able to transparently (dis)allow the service providers to access their data and control the granularity of the access (full, anonymously). They should be able to give permission for applications to access the relevant attributes about them that will enable the right decisions to be made about their eligibility for benefits or the most appropriate treatment for any health conditions and to ensure their control over content that they have created, while avoiding the need to link that data with their personal identity.

MIM4 is dealt with in detail in Chapter 9.

7.5.5 MIM5 fair and transparent AI

The aim of setting up a local digital ecosystem is to bring together information from many areas of city life to help ensure that the city can be managed more effectively and more focused around the needs of the citizen.

AI and algorithms will have a key role in making sense of that data and some of those algorithms will be decision-making. It is therefore vital that the algorithms that use that data are fair and transparent and that they use appropriate data from the data ecosystem appropriately to make decisions.

[10] https://github.com/FIWARE-TMForum/Business-API-Ecosystem

Cities need to be able to test whether products and services they are procuring are fair, trustworthy, and transparent and to ensure the appropriateness and accuracy of data used both in training the algorithmic systems as well as used by those systems in decision-making.

MIM5 will provide the technical capabilities required to check that the algorithmic systems offered by suppliers comply with the requirements for fairness, trustworthiness, and transparency through identifying or developing a relevant set of APIs.

7.5.6 MIM7 geospatial information management

MIM7 aims to provide minimal interoperability mechanisms related to geo-temporal data. However, there are many existing geo-temporal data standards that are of relevance to cities and to propose the full list would not be compatible with the concept of MIMs. MIM7 is therefore being developed as a number of parts.

The discovery, querying, retrieval, visualisation, and editing of geospatial information based on location and temporal criteria can be achieved through open standard formats, protocols, and preferably through the use of standardised API interfaces. Integrating context information with geospatial information can be enabled by the context management API and geospatial management API through common data information models defined in the MIM2 data models.

The minimal requirements to be included in MIM7 will enable access to the data that is necessary to enable the above to be done.

7.6 MIMs Plus

The European Commission is supporting the development of a specifically European version of the MIMs known as MIMs Plus to help fulfil the aspirations captured in the Living-in.EU declaration.[11] MIMs Plus is based on the existing minimal interoperability mechanisms plus some additional fundamental building blocks – hence the name: MIMs Plus.[12] An operational guidance paper is also being developed with practical guidance on how the specifications captured in the MIMs can be used in practice.

[11] https://www.living-in.eu/declaration

[12] Latest version available at https://www.living-in.eu/mimsplus

MIMs are aimed at providing consistent global processes to enable a global market, while MIMs Plus is aimed at setting these in the European Policy landscape. For the time being, the only difference is that the MIMs Plus is described in a document that sets them in that policy context. However, at some stage, there may need to be somewhat different technical specifications – for instance, when the European Digital Identity Framework and Digital Wallet are implemented, this will impact on MIM4 personal data management and the technical requirements for MIM4 in Europe will need to be slightly different to the global requirements.

8

Health Data in a Smart City

Vincent Keunen

Andaman7, Belgium
Email: vincent.keunen@andaman7.com

Abstract

Smart cities are cities where healthy living is encouraged. And to improve health, all healthcare professionals must agree to share the data they collect with patients. Interoperability is critical and we describe below where there is a conflict of interest and discuss anonymisation and pseudonymisation and the need for privacy and security. We describe the progress made in the USA and how Europe and smart cities could benefit from that experience. We dive into the specifics of health data and why a distributed approach is favoured by many.

8.1 Is Health Data Important for a Smart City?

Smart city and health data – what is the link? Well, if we go back to the fundamental roles of "city managers", they really are in charge of organising the life in the city, so that inhabitants live well together. And, clearly, healthy citizens will have a better life; so making sure they are healthy is important. Besides providing good care, this is also about empowering patients with their data so that they can receive better care and optionally contribute to research. So the "health data infrastructure" of a city is arguably as much important (or even more?) than other types of data.

But we have a conflict of interest.

8.2 The Conflict of Interest

In the health sector, there is a clear conflict of interest between the individual and the "common good".

DOI: 10.1201/9781003399384-9

Individuals want – and do have a right to – privacy. Without privacy, risks arise: the risk that you will not find a job if your employer is concerned with your health. The risk of paying higher premiums because you are at a higher risk, defeating the "solidarity" aspect of insurance, where everyone pays a little bit to cover the large expenses that some of us will face in case of serious health problems. The risk of reputation – some of us, with a public life may want to be discrete on conditions we are facing.

But for the common good, that is for finding new treatments, for improving our health systems (operationally and on the cost aspect), and for possibly other reasons, we need large quantities of data on whole populations. And that is in direct conflict with the needs of individuals.

8.3 Maybe Anonymisation is a Solution?

Is anonymisation a solution?

Anonymisation is the process of removing all PID (person identifying data) from a dataset. That can help avoid having access to one specific person's data. But, unfortunately, this is not sufficient. For example, it is sometimes needed to link data for the same person from various sources. For example, to have a long-term (longitudinal) view on the person's health – what condition they developed, when, and what is the link with their life habits, food and drugs consumption, etc. Full anonymisation is not a solution in these situations.

Pseudonymisation can help, however. What is this? It is the process of assigning a unique identifier to a given person who does not allow the viewer of the data to go back and find the identity of the person. Cryptographic techniques do exist to create a pseudonym from the identity of the person without a reverse process being reasonably available (or at least, it would take a huge amount of time with current computing resources to calculate the original PID).

It is still a risky process because some researchers have shown that, even with anonymised/pseudonymised data, under some circumstances and with additional identified data cross-linking, there are sometimes ways to re-identify anonymous data. See "Data re-identification" on Wikipedia for more details.[1]

Clearly, technological approaches to protecting citizens are only part of the solution. We need strong – and modern – laws to complement technology. It must be a big risk, with financial penalties, for companies and other institutions to use data in ways that could hurt citizens. The challenge here, as always, is that technology advances at a much faster pace than law does.

---

[1] https://en.wikipedia.org/wiki/Data_re-identification

GDPR is an excellent step in the right direction. And, surprisingly, many non-EU states are moving in the same direction as what EU has proposed (see the California Consumer Privacy Act,[2] several other US states privacy acts, Brazil's LGPD,[3] and many others).

8.4 Health of Citizens and Health of the City

So if a smart city needs to be a healthy city, many aspects must be considered. The environment, infrastructure, and ecological approaches to managing the city are important and are usually a focus of city managers.

But when it comes to the health of citizens, the whole healthcare ecosystem must be considered and that is sometimes less of a concern for cities. By "healthcare ecosystem", we mean hospitals and clinics – with their specialists, labs, general practitioners, nurses at home, physiotherapists, midwives, and all healthcare professionals – and maybe also caregivers, whether they are part of the family or paid professionals.

All of these players must have a role in the city, and appropriate financing mechanisms. Sometimes, cities may help incentivise some roles if they are missing or underrepresented.

Medical research is also a focus that some cities or regions could and should consider. One of the goals of medical research is to find new treatments that are effective and safe. The medical research field is already well developed in many regions of the world with pharmaceutical companies, medical device manufacturers, and bio-techs as sponsors of such research.

But cities have a role to play because some diseases are specific to some areas (malaria is only present in some regions of the world) or some ethnic groups. And it is recognised that, today, we need more diversity in medical research to improve treatments for some groups or areas.

Cities could contribute to medical research by helping the sector recruit patients and promote data interoperability in its own ecosystem. Interoperability is also critical to good care and control of costs.

8.5 Health Data Interoperability

Health data interoperability is still a very big problem in Europe today, as is in many areas of the world. However, some countries do progress significantly, like the USA (see below).

---

[2] https://en.wikipedia.org/wiki/California_Consumer_Privacy_Act

[3] https://en.wikipedia.org/wiki/General_Personal_Data_Protection_Law

8.5.1 Why is it hard?

Why is interoperability in health so problematic while, for example, financial institutions have solved the "financial data interoperability" problem for a long time? There are good and bad reasons for this.

Amongst the good reasons are the above-mentioned risks for privacy, for the right to be forgotten and other important elements for the citizen, for the individual. Security is a direct consequence of this: health data security is crucial to respect the privacy of patients.

Another good reason is that health data is extremely diverse. We talk about health data – versus medical data – because we need to manage – in addition to medical data from a doctor, hospital, or any healthcare practitioner (HCP) – various types of wellness, lifestyle, activity, nutrition, sleep, genetic, occupational, and medical research data (refer to Figure 8.1). Some use the more scientific term "omics" for this.[4]

Health data is very diverse and the needs vary (identified vs. anonymised data for example).

8.5.2 Unstructured data

Unstructured data includes text documents, medical images, pictures of text documents, etc., in many formats (for text documents: TXT, CSV, PDF, RTF, HTML, Word, XML (some), HL7 CDA (most), KMEHR, etc.; for images: jpg, png, gif, bmp, svg, XML, HL7 CDA, etc.).

Text documents, as such, can currently mainly be used to present the information to human readers (and transmit it). Through advanced services, that unstructured data could be converted to structured, codified data via NLP/ML (natural language processing/machine learning) systems.

8.5.3 Structured data

Structured data contain either numeric values (in many units and unit systems – imperial or metric) and codified values. There are also several codification systems: oftentimes custom but also more standard codes like LOINC, SNOMED CT, ICD, etc.[5]

---

[4] https://en.wikipedia.org/wiki/Omics

[5] LOINC: https://loinc.org and https://en.wikipedia.org/wiki/LOINC
SNOMED CT: https://www.snomed.org/ https://en.wikipedia.org/wiki/Systematized_Nomenclature_of_Medicine
ICD: https://www.who.int/classifications/classification-of-diseases https://en.wikipedia.org/wiki/International_Classification_of_Diseases

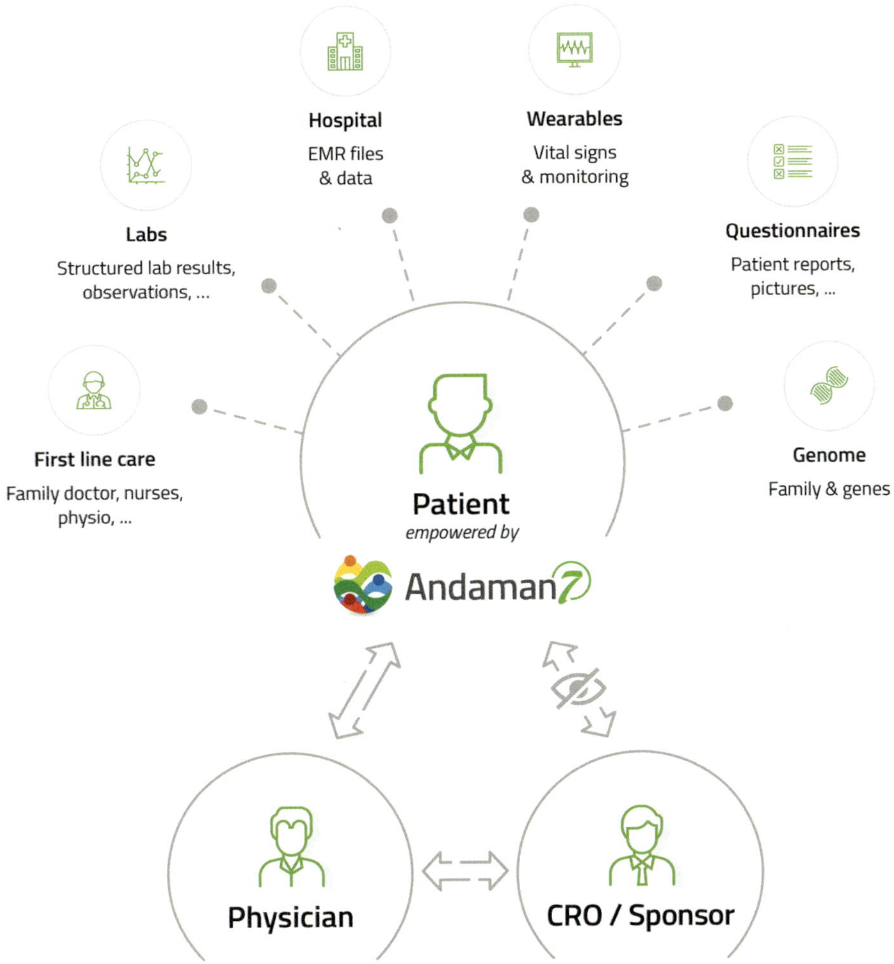

Figure 8.1 Variety of health data.

Structured data also include:

- numerous parameters coming from a growing set of wearables and connected devices, both consumer and medical grade;
- structured data resulting from rich questionnaires, trials, quality of life (QoL), ICHOM (International Consortium for Health Outcomes

Measurement), PRO (patient reported outcomes), RWE (Real World Evidence), etc.[6]

- data from Apple HealthKit on iOS devices and from Google Fit on Android devices.

But there are also less good reasons for the lack of interoperability like medical software lock-in by vendors, fears of transparency of some hospitals and healthcare professionals (revealing inefficiencies, excessive costs, or medical errors), and a simple resistance to change. But none of these serve the patients' interests and should be fought against.

8.5.4 Is the situation different in the USA?

The USA is a good example of significant progress on health data interoperability. A few years back, under the "meaningful use" initiative, all software vendors and users of these software tools were obliged to offer "standard access APIs" which means standard ways of accessing the data by external parties (other companies, patient's representatives, other institutions, etc.). Modern and mostly well-accepted standards like FHIR, LOINC, and SNOMED were chosen to improve interoperability.[7] This took a few years to accomplish.

Unfortunately, after these efforts, many hospitals still resisted sharing their data with patients. In early 2021, several patient associations made it clear to the US government that there were many impediments to them for accessing their data and the government passed new laws to impose financial penalties to these organisations (see "Denying Patients Access to Health Records/Exceeding Timescale for Providing Access" at https://www.hipaa-journal.com/common-hipaa-violations). Now, the situation is finally getting better for patients in the United States and hopefully Europe and smart cities may draw lessons from their experience.

---

[6] QoL: https://en.wikipedia.org/wiki/Quality_of_life
ICHOM: https://www.ichom.org/
PRO: https://en.wikipedia.org/wiki/Patient-reported_outcome; RWE: https://en.wikipedia.org/wiki/Real_world_evidence and https://www.fda.gov/science-research/science-and-research-special-topics/real-world-evidence

[7] FHIR: https://hl7.org/fhir/ and https://en.wikipedia.org/wiki/Fast_Healthcare_Interoperability_Resources
LOINC: https://loinc.org and https://en.wikipedia.org/wiki/LOINC
SNOMED CT: https://www.snomed.org/ and https://en.wikipedia.org/wiki/Systematized_Nomenclature_of_Medicine

8.6 The InteropEHRate Project

To help improve the situation in the EU, the InteropEHRate project, described in Chapter 2, has developed a number of protocols to improve cross-border interoperability.[8]

The project enables patients to be in full control of the usage and the routes of their health data. The central instrument, being laid in "patients' hands", is the smart EHR (S-EHR), leveraging a set of new protocols for secure and cross-border exchange of health data. Andaman7 is the reference implementation for the S-EHR.

8.7 Data Ownership and the Distributed Approach

Besides interoperability, there is also a lot of discussion on "data ownership". Is data generated by a hospital or a healthcare professional their property? It is especially critical because that data is about an individual person and their health. It is only ethical that pretended data ownership does not interfere with a person's good health. Retention or not sharing data can have dramatic impacts on patients, greatly reducing the quality of care or its cost. The healthcare industry should never abuse their power in ways that could hurt patients. Patients are citizens in a fragile moment of their existence. Healthcare is not and should never be an industry like any other. Ethics must play an even more important role than in the general industry.

At the same time, we should not forget that the patient is paying the care provider – directly or indirectly via their insurance or social security. It is only natural that, as a result of this service, patients receive the information and data associated with their diagnosis and treatment.

Data ownership is not the right question, actually, and European laws for protecting patients and citizens move in that direction. Data is not a physical element. It can be easily (and with almost no cost) duplicated. Access to data can be given and access is becoming the crux of the question. It does not matter that much who "owns" a piece of data, but it is important to define "access rules". And GDPR is clear: patients should have access to their data, be able to correct them, have a right to be forgotten, etc.

So patients should have easy access to their health data – whoever is the "owner". And it should be accessible in a FAIR format. FAIR data are data which meet principles of findability, accessibility, interoperability, and

[8] https://www.interopehrate.eu
http://www.andaman7.com

reusability.[9] That means that patients should have access to the highest quality data and in a structured and/or codified format if they are available at the source. It is not good enough to give paper copies to patients. And it is not good enough to give them a PDF report from their lab results (or from any other data type). Both the PDF (for readability, comments, validation information, suggestions, etc.) and the underlying structured data must be provided (for example, in the FHIR format, with an LOINC codification). The structured data can then be processed on the PHR of patients. For example, with structured data, a patient can follow their PSA values over the years to control the evolution and the risk of prostate cancer.

Where should that data be stored? In the hospital? On national servers? Here, again, there have been many discussions over the years. A small number of countries (Denmark, Estonia) have implemented national systems with a central storage of all health data. This has the benefit of having a well-organised system, with advanced capabilities to process data for care and cost optimisation but also for research (also known as "secondary use of data").

However, many countries resist centralisation of health data, for fears of malevolent use of the data, and fears of too much information being available to the government – and the hackers that would succeed in breaking into those centralised systems.

Taking this into account, some stakeholders have proposed a distributed approach with no master/slave architecture.[10] This is a bit more challenging technically, but in the age of artificial intelligence and machine learning, it is actually very accessible. In the distributed architecture, data is copied to every location it is needed. The exchange of data is based on a peer-to-peer approach. With advanced traceability techniques, it is actually possible to have copies of data in various locations without redundancy or data conflicts. Data can be added to the EHR (electronic health record) of a given patient at any location, by anyone (who has the rights to do it), at any time. Data will then flow to the other locations it is needed.

This distributed approach is pretty novel, but it has been shown to be very effective. Patients using the PHR (personal health record) can collect their data from many sources, manually add information to it, and then share that data back to persons that are in their circle of trust. The traceability of the

[9] See https://en.wikipedia.org/wiki/FAIR_data for more details

[10] (see Andaman7 at http://www.andaman7.com

data is then used by healthcare professionals to give varying degrees of credit to the information at hand, depending on the source of the data.

The distributed approach is the only one to be future proof. Each stakeholder needs to stop thinking that they are the centre of the world and have their IT systems built in this way. We now live in ecosystems where several players can contribute to the health of citizens.

9

Personal Data Management and MIM4

Michael Mulquin

MIMs Ambassador, Open & Agile Smart Cities, Belgium
Email: michael@oascities.org

Abstract

The chapter starts by listing some of the many initiatives that aim to put the citizen in charge of how data about them is used. It points out that the large number of these initiatives demonstrates the importance of this issue, but also that it has resulted in a fragmented marketplace with competing technical approaches and processes.

It then describes how Open & Agile Smart Cities is working with many of these initiatives to develop a minimal interoperability mechanism on personal data management (MIM4). This will identify an agreed set of capabilities to enable the different solutions to be compared, a common legal framework that all can sign up to, and a technical requirement that will enable "good enough" interoperability between them and thus help to bring consistency into the marketplace.

9.1 The Fragmented Marketplace

We have already seen how one of the key challenges in today's world is putting the citizen back in control of their data. There are many initiatives that are working hard to tackle this issue. Solid and MyData have already been mentioned elsewhere. Others include the following.

- A NewGovernance (aNG):[1] An association promoting and enabling human-centric personal data sharing. The aim is to support and

[1] https://www.anewgovernance.org/

DOI: 10.1201/9781003399384-10

coordinate the many personal data sharing ecosystems that are being developed by developing standards and a governance framework. It has a strong European base but is firmly global in vision.

- Disposable identities:[2] This is an initiative pushing for the use of digital wallets to enable individuals to access services and take part in smart contracts without disclosing personally identifiable data.
- Privacy by design foundation:[3] It creates and maintains free and open-source software in which the privacy of the user is the most important. The most important application of the foundation is the identity platform IRMA, an acronym of I Reveal My Attributes. The foundation also aims to generally improve the development and usage of open, privacy-friendly, and well secured ICT.

There are also several European Projects such as DataVaults[4] and KRAKEN,[5] along with a number of other initiatives, which are each developing their own technical solutions to provide personal data management solutions and are primarily in the pilot or development phase. The aims of the different initiatives overlap but are not necessarily identical. Some projects focus just on personal data management, and others, such as Rennes Urban Data Interface (RUDI),[6] aim to support wider data sharing ecosystems but with personal data management being a key feature.

The large number of agencies and projects working in this area demonstrates the importance of the issue. However, it has led to a fragmented marketplace, with many different technical solutions and business models.

MIM4 aims to tackle this issue by identifying the key capabilities required and developing points of interoperability between the different solutions to help build confidence and support implementation.

9.2 MIM4

MIM4 focuses on personal data management (PDM), in other words, how to provide easy to use methods for citizens/users to control which datasets/attributes they want to share with solution, application, or service

[2] https://disposableidentities.eu/

[3] https://privacybydesign.foundation/en/

[4] https://www.datavaults.eu/

[5] https://www.krakenh2020.eu/the_project/overview

[6] https://uia-initiative.eu/en/uia-cities/rennes-metropole

providers under transparent circumstances, enabling trust between the different parties.

Specifically, it will provide technical and other guidance to support cities and communities to put in place the products and services that will enable their citizens to be in control of their personal data within the local data ecosystem. It will do this in a way that will make it easy for them to integrate their services with whatever credible personal data management systems their citizens may wish to use.

MIM4 will define:

1. the capabilities that cities and communities need to put in place to enable citizens to have control of their data within the local data ecosystem;
2. the requirements to enable "good enough" interoperability between existing services and projects that offer solutions for personal data management.

The work will include reviewing how MIM4 can be integrated with the other MIMs to support the effective personal data management within a local data ecosystem.

MIM4 will also point to sets of recommended solutions that will enable cities and communities to comply with these requirements.

9.2.1 Capabilities

MIM4 will address needs and requirements from two perspectives:

- that of individual citizens in terms of transparency and privacy preferences collection;
- cities and data using services (data controller/processors) in terms of authorisation and data usage control and enforcement.

The provisional sets of capabilities identified so far are listed below.

For individual citizens:

1. Citizens need to be able to choose the operator they wish to manage their data and to move from operator to operator.
2. Citizens should be able to access their data through many different channels.
3. Citizens should be able to use the identity of their choosing, in best cases a keychain of identities can be defined, so that users can choose the identity per service.

4. Citizens should have insight into what personal data is available, stored, shared, etc., by the providers of the applications and/or services they use.
5. Citizens should be able to request changes to or deletion of part or all personal data available, stored, shared, etc., by the provider of the applications and/or services in use. The providers would need to comply with these requests unless there were legally justifiable reasons not to do so.[7]
6. Citizens should be able to indicate in which circumstances what personal data is "free" to use for which parties through a "permission arrangement".
7. Citizens should be able to grant consent to providers of the applications and/or services, be it governmental or businesses, that attribute-based, decentralised storage, and "revealing" of personal data attributes provides full service and access to these applications and/or services.
8. Citizens should be able to roam with their data between cities and internationally.

For cities and data using services:

1. Cities need to enable users to handle consent, allow and revoke access, and have full transparency on their personal data.
2. Permission management needs to be handled preferably on the attribute level. Personal data processing should be described in a fine-grained manner, by covering all aspects (purposes, processing, types of data, etc.) in a standardised manner (see, for example, W3C Data Privacy Vocabulary[8]).
3. Personal data management needs to have an open API in line with MIM1 to broker data and standard data models MIM2. Data sources need to be open and documented, and discoverable via MIM1, listing their data via MIM2. Operators may benefit from being groupable at joint initiative of cities with close ties.

[7] For instance, the citizen cannot expect information regarding their age or any other key factual piece of information to be changed so as to be incorrect, specifically in a way that will affect their eligibility for services.

[8] https://dpvcg.github.io/dpv/

4. PDM systems need to manage the personal data to a high level of security (the detail of how to do this will be dealt with by MIM6).

5. PDM systems need to be flexible enough to handle methodologies that require personal data pods to store the data as well as those that utilise personal data spaces or that allow the data to continue to be stored by the relevant organisation, but where the subject of the data is able to exercise rights as to its use.

This list of capabilities will be tested with the various projects and initiatives working in the field to develop a consensus-based set that can be used to review and compare different initiatives that aim to enable personal data management.

9.2.2 Requirements

A detailed proposal for interoperability between personal data management operators has been reviewed in detail by MyData Global, Vastuu Group, Forum Virium Helsinki, RUDI (the Urban Data Initiative of the city of Rennes), the DataVaults, DataPorts and KRAKEN European Projects focusing on personal data management, and the CAPE personal data management solution developed by the engineering group.

This proposal has two pillars:

- Pillar 1: One connector for all personal data management operators.
- Pillar 2: Legal framework governance.

The proposal is described in the paper "Towards Interoperable Personal Data Management within Smart Cities: Minimum Interoperability Mechanism 4" that can be accessed at: https://mims.oascities.org/mims/oasc-mim4-trust/references

Effectively, this defines a connector that enables any personal data management provider that complies with the legal agreement to be able to access data from any data source that is MIM4 compliant. In this way, each personal data management provider can innovate freely around their technical solution, provided that it enables the capabilities defined in MIM4, while data providers only need to provide a single method for them to access the data.

This review indicated that the proposed interoperability mechanism is a feasible way of enabling a level of interoperability between all of these and is likely to be relevant to all personal data management solutions. All the above initiatives have also agreed to work together over to develop demos to test the

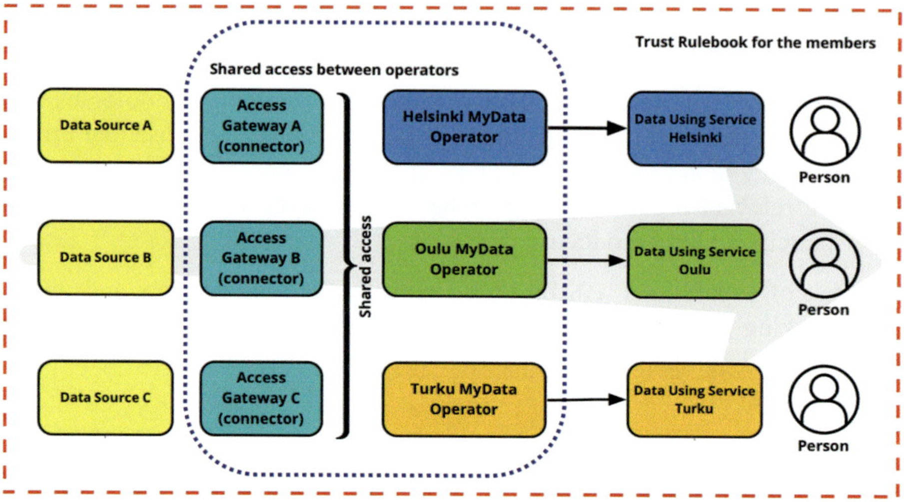

Figure 9.1 Shared access.

technical suitability of this solution in practice. In addition, the Flanders Data Utility Company[9] that is implementing a Solid-based approach to enabling the citizens of Flanders to manage their personal data has agreed to join the testing programme.

The organisations taking part will also review the proposed legal agreement to ensure that it is practical and covers all the key issues.

9.3 The Link with National ID/Citizen Cards

Within Europe, the planned European Digital Identity Framework being developed under eIDAS will make it possible for every person eligible for a Member State national ID card to have a digital identity that is recognised anywhere in the EU. It will provide a simple and safe way to control how much information an individual wants to share with services that require sharing of information. It will operate via the use of digital wallets available on mobile phone apps and other devices to:

- identify the citizen both online and offline;
- store and exchange information provided by governments, e.g., name, surname, date of birth, and nationality;

[9] https://www.vlaanderen.be/digitaal-vlaanderen/het-vlaams-datanutsbedrijf/the-flemish-data-utility-company

- store and exchange any other information provided by trusted private sources;
- use that information to demonstrate the right of the individual to access services.

Specifically, for Europe, the development of the European Digital Identify Framework and Digital Wallet will have a significant impact when it is launched in the next two to three years and the MIMs Plus version of MIM4 will need to take that into consideration.

More widely, the European Digital Identity Framework not only builds on multi-purpose National Identity Cards within member states, for instance Estonia, but on the work many cities around the world are doing to develop citizen cards to help citizens access local services. It is also worth noting the White Label citizen card solution being developed by Eurocities. These citizen cards may be provided using a dedicated smart card or via a smartphone app.

MIM4 will be developed in a way that will align with these initiatives.

10

Standards for Citizens

John Ketchell

European Association for Coordinating Consumer Representation in Standardisation (ANEC)
Email: john@raisingstandards.eu

Abstract

This chapter faces the theme of standardisation oriented to smart communities. This represents a great challenge for all the standardisation bodies in the EU as it needs to immerge technical ICT-based approaches into a civic and social based framework and raise acknowledgment on the importance of such an issue by all the involved parties, city administrations, and citizens.

The chapter provides the background on standardisation activities carried out by European bodies and provides some insights on current work in the field of citizens' standards in smart communities, in terms of both ongoing initiatives and failings faced. Eventually, a view on further needed actions is provided to enhance citizen services, including the data aspects, citizen-oriented management of local authorities, and citizens' security, both physical and online.

10.1 Introduction

Like every other aspect of information and communication technologies affecting our daily lives, smart city and community implementation requires a modicum of standardisation, whether this comes from a large multi-national imposing its own "standardised" solutions, from consortia of like-minded industrial companies devising solutions for interoperability or from formal international or European standards with a wide consensus.

Standardisation in the smart community domain is somewhat in its infancy. It mainly seeks to address either the technical requirements for

DOI: 10.1201/9781003399384-11

communities to carry out their processes or how a community manages these. For various understandable reasons, these solutions do not yet normally take proper account of significant aspects taken from the perspective of the citizen in such a community.

Nevertheless, the smart community concept offers considerable opportunity, not only for citizens to have an improved living environment in which they can benefit from effective services but also for them to influence matters affecting their daily lives. At the same time, we need to ensure equality of treatment for all citizens, and account needs to be taken of data privacy concerns relating to their personal information. So far, the needs of the citizens themselves have been somewhat set aside in the debates concerning smart communities, but, more than ever in a post-coronavirus world, it is important to keep these needs at the forefront of societal development.

This will require increased cooperation, and, indeed, it strongly reinforces the need for standardisation. Of course, some standardisation activities not specifically aimed at local communities as such, for example, the standards in support of data protection and privacy, will nevertheless benefit them. But, equally, it will be necessary to ensure that the development of these wide-reaching standards takes proper account, not simply of the needs of a local authority but those of its citizens.

10.2 The Background

There is a regulatory framework to standards-making at European level. For many years, legislation has set forth the overall concept of standardisation within the European system, and given due recognition to the three European Standardization Organisations, CEN,[1] CENELEC,[2] and ETSI.[3] Many of the standards of interest to the regulator are described as "harmonised" European standards and are drafted in support of legislation. However, these are of relatively small importance in a smart community context – the bulk of activities within the European organisations is purely "voluntary", in that the standards (and other documents produced by the consensus of participants) are drawn up on the basis of the consensus of participants.

European Standards, nevertheless, undergo a formal process of open comment and national voting (in the EU, the EEA and some additional countries, to a total of 34, or more in the case of ETSI) prior to their publication,

[1] www.cen.eu

[2] www.cenelec.eu

[3] www.etsi.org

and conflicting standards which may exist at national level must be withdrawn. But, in addition, the standards organisations publish many documents that reflect simply the participants' consensus without that formality.

The current overall framework is described in a European Regulation (EU) 1025/2012. This regulation imposed an additional set of obligations on the European standards process to take improved account of the needs of societal stakeholders and SMEs. However, it is difficult for these entities at an individual level to drum up the resources to send a lot of representatives to standards meetings, even to find the time to participate in electronic meetings.

The European Commission and EFTA have, therefore, provided support to four European-level associations representing multiple stakeholders to participate in the European standards process. The result is the existence of the "Annex III" organisations (so-called from the relevant Annex to Regulation 1025). One represents SMEs, and the others act as advocates for the particular aspects of civil society – environmental organisations, organised labour, and consumers.

Thus, the organisation representing consumers, ANEC ("The European Consumer Voice in Standardisation")[4] has as its member's national consumer bodies in many European countries. As only one of many activities, ANEC is seeking to improve consumers' (i.e., citizens') contribution to standards for smart cities and communities.

Early in 2022, the European Commission adopted a new European standards strategy, and this lays even more emphasis on the urgency for the standards organisations to take account of the needs of all interested parties and, therefore, to improve participation from civil society and SMEs. In response, the European Standards Organisations are considering how they can be more inclusive, given that these organisations lack expertise and indeed resource to be fully participating, except in the most niche and specialist areas.

Thus, for example, a sub-group of the ETSI Board is examining during 2022 and 2023 the ways to improve processes and procedures with an inclusive approach, as well as how to involve potential end-users – including local communities and citizens themselves – with some worthwhile contributions to make, but who are not, for whatever reason, currently adequately involved.

10.3 Citizen Standards in Smart Communities

Several years previously, the standards organisations had been seeking to address the smart community aspects relating to citizens.

[4] www.anec.eu

Thus, at the European level, a co-ordination Group of CEN, CENELEC, and ETSI had proposed the development of a technical report[5] on these citizen-related issues, and this proposal was later taken up in the European Commission ICT Standardisation Rolling Action Plans for 2016 and 2017.

This report – not a standard but a narrative setting out what is required – has sought to clarify what further standardisation is needed on citizen issues in the context of smart cities (e.g., standards on what, where, when, etc.), while taking full account of other relevant standards activities under way.

The document – ETSI Technical Report 103 455 – is available free of charge to download.

The report cites a number of serious issues specific to smart cities standardisation, before reaching a number of recommendations on how to try to overcome these and on particular standards activities that are desirable in the interests of the citizens.

On the first, "organisational" aspects:

- cities do not know standardisation and are confused by the plethora of standards and indeed committees with complex rules;
- cities find it very difficult or impossible to participate;
- funding models for standards in this area are inadequate;
- services are not designed for or even accessible to citizens;
- protection for citizens' data protection and security may be inadequate.

In order to address these failings, the report suggests better financial support for the participation of cities in standards activities, perhaps along similar lines to an existing scheme under Horizon Europe to support participation of experts in international ICT standards activities. In terms of cities' understanding of standardisation, the report recommends a better engagement between standards organisations at national level and local government structures, to improve policy-makers' understanding of the importance and relevance of standardisation in their universe. Such engagement can also result in a better process for taking smart community views into account in the standards process to reflect the difficulties local authorities may have to participate.

Turning to the actual standardisation recommendations, these have been divided into three broad categories. First, there is an acute need for

---

[5] https://www.etsi.org/deliver/etsi_tr/103400_103499/103455/01.01.01_60/tr_103455v010101p.pdf

consensus-based guidance material, to help smart communities engage better with citizens' needs. Examples include material for training local authority personnel on citizen-related issues, how best to protect citizens' security (both physical and electronic), and how to ensure data protection.

This need can extend in, perhaps, a more binding way to the second category, codes of conduct, or good practice to be observed, albeit on a voluntary basis. This might include citizen-oriented complaint and redress procedures, perhaps along the lines of the existing European On-line Dispute Recognition arrangements or of a good practice in management of services for the individual citizen.

Last but not least, there is the need for more formalised standards. These should cover how services should be designed to meet the needs and capabilities of the citizen, thus making these services user-friendly but also accessible to the less privileged (bearing in mind also the requirements of the European Accessibility Act).

Other more specific citizen-oriented standards are needed for security and data protection, to supplement existing more generic efforts. For example, a standardised approach is needed to cover citizen uses for, and requirements from, the data spectrum. Shared data needs to be involved – for what purpose the data is being shared, for what purpose it is being used, and what security and access controls are required to meet both privacy-preserving and security-minded requirements for the citizen.

10.4 Looking Ahead

The implementation of these recommendations is challenging. Where there are immediate commercial incentives for standardisation, industry seldom has difficulties in ensuring participation in the interests of its markets. Clearly, the commercial incentives for smart city standardisation are limited, and, as noted above, into the bargain local communities' understanding of standardisation is rather poor.

The implementation of the recommendations of the ETSI Technical Report has been affected by these considerations, which, of course, have been exacerbated by the pandemic. The European Standardisation Organisations have agreed to package the recommendations into three broad projects, of which the highest priority has been given to citizen services, including the data aspects. A second project is intended to enhance citizen-oriented management of local authorities, and a third should examine citizens' security, both physical and online. CEN, the European Standards Committee, has started a new Technical Committee (identified as TC465) on Sustainable and Smart Cities and Communities, and this has now adopted a first work item on citizen services, which will be in the form of an overall work programme proposal, noting that this new field for standardisation needs a number of separate and parallel actions. This activity will start – hopefully with a good involvement of representatives of local communities – as soon as the resources are available.

However, this Committee must not work in isolation. It needs to be linked to a wide range of other generic standards initiatives, and also to regulatory needs, for example, those stemming from the EU Data Governance Act and the future Data Act. The standards produced need to privilege the rights of the individual citizen and provide a non-burdensome set of recommendations to be easily and fruitfully fulfilled by all involved parties.

11

Business Models

Shaun Topham

Lead, EU Smart Cities Marketplace, "Citizen's Control of Data" Initiative
Email: shaun.topham@eu-forum.org

Abstract

This discussion will be structured by first considering the issue of business plans at the "smart city level", before focusing on the business models associated with urban data platforms and before delving into the issues faced by adding a citizen's personal data into the mix, and how this addition can help contribute to the evolution of urban data platforms.

However, this is not simple, as we are considering an amalgam of interrelated and interconnected social, human, technical, and economic variables, all at the same time and focused on an ever-shifting and ever-developing city landscape. The situation could be described as "the scourge of tidy minds!"

Approaches derived from a series of EU funded projects will be described.

What is significant is the amount of information which collectively these projects featured have produced on this topic and derived from hundreds of sources in their research. Many provide a range of tools and advice far in what you would expect to find on a project website.

11.1 Introduction

This discussion will be structured by first considering the issue of business plans at the "smart city level", before focusing on the business models associated with urban data platforms and before delving into the issues faced by adding a citizen's personal data into the mix, providing a stimulus to the deployment of urban data platforms.

DOI: 10.1201/9781003399384-12

However, this is not simple, as we are considering an amalgam of interrelated and interconnected social, human, technical, and economic variables, all at the same time and focused on an ever-shifting and ever-developing city landscape. The situation could be described as "the scourge of tidy minds!"

A potentially useful tool to help with the analysis of "How a smart city is run" is a technique developed by the UK Open University in relation to working out how best to describe "How a country is run".[1]

To answer this question, three common models are used simultaneously, like shining a light on an object with three different torches. The three models being hierarchy, markets, and networks defining them loosely as follows:

- A hierarchy is as how we commonly use it, with a government or corporate organisation being able to pass down instructions so that something happens.
- The key concepts in market coordination are enforceable contracts and property rights. Communication in a market is through prices, which serve as the basis for behaviour of both buyers and sellers.
- Networks are characterised by the complementarity of member's interests and the reaching of an agreement between them. Coordination and agreement within networks are achieved through discussion rather than through the impersonal mechanisms which work in markets and hierarchies.

The value in the use of these three models to analyse the process of coordinating smart city activities and usage of data is that a different perspective is reached through each one, which, when combined, give a better view of what is happening.

On one end of the scale is the logic of the UK post-Brexit advocates of the free market approach. A very serious argument has to be put forward to the government for any form of hierarchical government intervention in the data economy, despite the numerous reasons why data cannot be classified as a simple product, conforming to a theoretical concept of how a market functions. This is reflected in the current strategy prepared for the Department for Culture and Media.[2]

As we recognise, data can be used repeatedly and in different combinations and for a variety of reasons, in addition to those for which it was collected or

[1] G. Thompson, J. Frances, R. Leva£ic, and J. Mitchell, Eds., MARKETS, HIERARCHIES AND NETWORKS. The Coordination of Social Life. Newbury Park: SAGE Publications, 1991.

[2] "Increasingaccesstodataheldacrosstheeconomy-GOV.UK."https://www.gov.uk/government/publications/increasing-access-to-data-held-across-the-economy (accessed Jul. 26, 2022).

those which might make a profit for an enterprise. It is a non-rivalrous good. And as we touch on in later chapters, the price mechanism for data cannot rely on simple supply and demand graphs and a whole new approach is sought after to put value on what is occurring within a complex eco-system, lubricated by interlocking networks.

And, of course, a government at any level, with wider aspirations than "rolling back the state" has a wide range of policy objectives which can be supported through a more enlightened approach to data and in the light of it being a "non-rival" good available for use by many at the same time, without it diminishing. This is particularly so, when faced with the overall objective of becoming climate-neutral.

We will look further how we put a value on data in Chapter 13. What is clear is that a "one size fits all" solution will not emerge.

11.2 Business Models and Smart Cities

When the European Innovation Partnership on Smart Cities and Communities was launched a decade ago, it was to primarily create a number of "Lighthouse Initiatives" to deliver common smart city solutions. This would enable the creation at scale; whilst reducing risk for political decision-makers as well as investors", the Lighthouse Initiatives were expected also to adhere to the fulfilment of its second objective: to "Apply new business and financial models, public-private partnerships that combine industry with public investments at European, national, regional and local level", in order to deliver improvements faster across the three vertical areas identified as priorities".[3]

The three priorities are sustainable urban mobility/sustainable districts, built environment and integrated infrastructures, and processes across energy, ICT, and transport.

Funded Lighthouse projects were expected to contribute and to come up with novel business models to close potential gaps between achieving the policy goals and the availability of investment, which should become replicable and adopted widely within the smart city movement.[4]

[3] "EIP-SCC Strategic Implementation Plan - Smart Cities Marketplace." https://smart-cities-marketplace.ec.europa.eu/media/2261 (accessed Jul. 26, 2022).

[4] Smart Cities, "Towards A Joint Investment Programme For European Smart Cities. A Consultation Paper to Stimulate Action," Marketplace of the European Innovation Partnership On Smart Cities and Communities, 2016, Accessed: Jul. 26, 2022. [Online]. Available: https://smart-cities-marketplace.ec.europa.eu/sites/default/files/2021-04/EIP-SCC_TOWARDS%20A%20JOINT%20INVESTMENT-Paper.pdf

The following sections will trace the evolution of the approach to the smart city business model followed by the Lighthouse projects.

11.3 Smart City Networks Creating Best Practice Repositories

Thus, in recent years, a plethora of projects and cities have been working on novel business cases at a smart city level. The Lighthouse cities themselves have formed a network to continue to collaborate within scalable cities, as part of the Smart Cities Marketplace.[5]

Another initiative within the Smart Cities Marketplace is "City Wisdom" which forms a knowledge hub for business models and finance from real-world smart city projects which has the intention to "Identify and collect a complete collection of business models and finance analysis from city projects".[6]

11.4 SmartEnCity Project

Along with other Lighthouse projects, work was carried out in developing a new business model to identify the path to be followed on project replication and on business models that will make projects similar to SmartEnCity economically feasible, without public subsidies. They used the basic business model canvas as a basis for developing their business models.[7]

Recommendations are based on existing business models being implemented for working areas such as building retrofitting; energy supply and use; smart mobility and investments in ICT in Lighthouse projects. Here, a logic can be seen in using the "normal" business model canvas as the individual application areas appeared as being suited to being approached almost as distinct products.[3]

---

[5] "Scalable Cities - Smart Cities Marketplace." https://smart-cities-marketplace.ec.europa.eu/scalable-cities (accessed Jul. 26, 2022).

[6] "City Wisdom - Smart Cities Marketplace." https://smart-cities-marketplace.ec.europa.eu/action-clusters-and-initiatives/action-clusters/business-models-finance/city-wisdom (accessed Jul. 26, 2022).

[7] "SmartEnCity -D2.3. New business models, procurement schemes and financing mechanism for the smart city projects ," SmartEnCity Consortium, 2016, Accessed: Jul. 26, 2022. [Online]. Available: https://smartencity.eu/media/smartencity_d2.3_new_business_models_procurement_schemes_and_financing_mechanisms_for_smart_city_projects_v1.0_1.pdf

The SmartEnCity approach was that from the European Innovation Partnership on Smart Cities and Communities Strategic Implementation Plan. This covered the issues faced and the approach to business models at the same time:

> "In most cases, new investments will be needed to generate the broad uptake of smart city solutions. However, due to the economic crisis and increased demand for public services (demographic change, care, transfer of tasks from central government levels etc.), the public sector – locally and centrally – has limited budgets. This means that new market-oriented and sustainable strategies of public private cooperation must be developed and cities must seek greater levels of external investment. The investment community seeks certainty, and scale. However, most cities, at an individual level, presently deliver neither of these. Continuing 'business as usual' will not create enough value and scale for city administrations, cities, businesses and solution providers".

The SmartEnCity project defined its business model according to the business canvas model.[8]

It was selected as a model that everybody can understand and can make themselves. "The challenge is that the concept must be simple, relevant, and intuitively understandable, while not oversimplifying the complexities of how enterprises function".

We will see the adaptation and evolution of this original canvas business model approach as a recurring theme as cities and projects struggle to define their business model approach.

Issues which SmartEnCity encountered included that the goals developed in the three vertical priority areas cannot be achieved using traditional methods, for several reasons.

- First, there is a need for smart solutions that are developed in collaboration between citizens, local and global industries, municipal utilities, and the local public agencies – this often defies conventional procurement and tendering procedures.
- Second, although solutions must be local, such typically small-scale individual solutions are unnecessarily expensive and preclude the development of a business case for innovative smart city solutions at pan-European scale.

---

[8] A. Osterwalder and Y. Pigneur, Business Model Generation. Strategyzer, 2009. Accessed: Jul. 26, 2022. [Online]. Available: https://www.strategyzer.com/books/business-model-generation

- Finally, the matching and combining of complex city needs with industrial needs for long-term process and product innovation can be improved significantly.

But a major obstruction identified, and one we will return to in Chapter 13, is that progress is hindered by the need for all stakeholders involved in the processes of city transformation to be able to identify the added value of the investments. This requires a deeper public and private collaboration and a stronger integration of the value chain, with it matching city needs with industrial solutions as a strategic issue to leverage public and private investments.

11.5 Urban Data Platforms

We started by looking at the smart city approach as a whole, but we begin to see the importance of factoring in the general use of data within a smart city.

So, the business models being considered for deploying smart cities also spilled out into the development of models for deploying "urban data platforms".

The SmartEnCity Deliverable 6.6[9] focuses on the value added services on top of their "City Information Open Platform". In this document, they focused on creating a toolbox to help build local value added services that match the context. This involves identifying capabilities, partners, and use-cases, after which use-cases should be evaluated, deployed, and results validated. Section 4.3 of D6.6, "Examples of value added services", provides an extensive list.

This section contains a table with the most significant value added services examples identified in the "state-of-the-art" and broken down into the main verticals and other domains.

11.6 REPLICATE Project

The conclusion from the REPLICATE Lighthouse project, however, was that work still needs to be done in this direction, as it was concluded that:

"Supported by the empirical evidence provided by the municipalities about the REPLICATE project interventions, rounds of data collection from the REPLICATE cities, and an extensive literature search, this Work Package has found that replicable business models, as interpreted in a narrow firm-oriented sense as exemplified by much of the Business Model Innovation

[9] "SmartEnCity - D6.6. Strategies for Value Added Services ," SmartEnCity Consortium, 2017, Accessed: Jul. 26, 2022. [Online]. Available: https://smartencity.eu/media/del6.6.pdf

(BMI) literature, are not by themselves a sufficient means of bridging this financing gap".[10]

Further it was recognised that "close cooperation is required to make smart cities a more attractive investment, from devising innovative new business models to aggregating demand to access economies of scale and pooling risk". This implies an increased emphasis on the network as a means of coordination.

The REPLICATE Report on the Business Models of the Lighthouse Cities presents a framework for analysing the business models of the Lighthouse cities involved in the project. This framework is called the city model canvas which can be downloaded.[10]

11.7 IRIS Project

Like REPLICATE, the Iris project produced a study which took forward the evolution of the basic business canvas and built upon it, the smart city business model canvas.

This study reviewed business model development frameworks and developed a practical tool to help cities assess business models by adapting components of the business model canvas (BMC) and adding new ones that operationalise the smart city dimensions. The smart city BMC (SC-BMC) proposed provides a practical framework that supports developing and communicating a more holistic and integrated view of a smart city business model. It also supports creatively innovating toward more sustainable value creation. As a framework, the SC-BMC bridges sustainable value creation for business model development and smart city innovation.

11.8 IRIS Study and the Smart City Business Model Canvas (SC-BMC)

An extensive study by IRIS shines further light on the topic.[11]

As smart, data-based value propositions can generate complementary benefits and attract private investment, smart city business models should

---

[10] "REPLICATE - D2.2 Report on the Business Models of the Lighthouse cities," REPLICATE Consortium, 2017, Accessed: Jul. 26, 2022. [Online]. Available: https://replicate-project.eu/wp-content/uploads/2018/09/REPLICATE_D2.2_Report-on-the-Business-Models-of-the-Lighthouse-Cities.pdf

[11] P. Giourka et al., "The Smart City Business Model Canvas—A Smart City Business Modeling Framework and Practical Tool," Energies 2019, Vol. 12, Page 4798, vol. 12, no. 24, p. 4798, Dec. 2019, doi: 10.3390/EN12244798

address the ownership of data and legal issues for data privacy in the design phase of the solutions deployed. An extra component was thus proposed to complement the value proposition in the SC-BMC, where the value of the data generated offers a multiplier effect in creating new business opportunities and increasing quality of life.

The SC-BMC builds on Osterwalder & Pigneur's original business model canvas, borrows from the BMC for mission-driven organisations and from the "triple bottom line" BMC,[12] and incorporates the importance of using a network-centric approach where the quintuple helix actors co-create a value and utilise data that adds value to the network of actors and other stakeholders such as new ventures.

This study emphasised that it is the network of actors – and not the city or one venture alone – that creates, delivers, and captures value. In that sense, while the BMC was aimed at helping individual firms to fit their BM to circumstances, customers, and markets, the SC-BMC was shifted to assist a group of actors aiming to deliver value to a more diverse set of end-users in a collaborative effort. Mapping out this network early on should precede the entire SC-BMC canvassing exercise. The development of this tool is based on a publication of Fruhwirth *et al.* (2020).[13]

The canvas can be found on the Iris website.[14]

But the study also found that "In the literature, however, there is a lack of a business model framework to determine clear paths to steady revenues for smart city projects encompassing the specific characteristics of each city".

But in a smart city context, value creation is shifted from products and technologies themselves to focusing directly on the verticals identified earlier, "to the services provided to users. Therefore, a service-dominant business model is prominent in the business modelling design and the value creation logic".[15] So the network-centric value proposition and a service-oriented business model canvas were used to describe the value created by the network for each actor separately in a smart city context.

---

[12] A. Joyce and R. L. Paquin, "The triple layered business model canvas: A tool to design more sustainable business models," Journal of Cleaner Production, vol. 135, pp. 1474–1486, Nov. 2016, doi: 10.1016/J.JCLEPRO.2016.06.067.

[13] M. Fruhwirth, G. Breitfuss, and V. Pammer-Schindler, "The Data Product Canvas - A Visual Collaborative Tool for Designing Data-Driven Business Models," BLED 2020 Proceedings, Jan. 2020, Accessed: Jul. 26, 2022. [Online]. Available: https://aisel.aisnet.org/bled2020/8

[14] "The Smart City Business Model Canvas – IRIS Smart Cities." https://irissmartcities.eu/smart-city-business-model-canvas/ (accessed Jul. 26, 2022).

[15] R. Lusch and S. Nambisan, "Service Innovation: A Service-Dominant Logic Perspective," MIS Quarterly, vol. 39, pp. 155–175, Jul. 2015, doi: 10.25300/MISQ/2015/39.1.07.

11.9 REPLICATE Project

The main objective is for the three lighthouse cities and the three follower cities to have, at the end of the project, fully optimised smart city business models.

This involved analysing the different models and possibilities, and the key elements that the three cities involved (San Sebastián, Florence, and Bristol) took into account to select and work with their models. The task provided a tool called city model canvas (CMC) to reflect how smart city solutions create and deliver value to their residents, again derived from the basic business model canvas.[8]

The main objective was to develop new sustainable and cost-effective services to citizens and public administrations providing integrated infrastructures that improve efficiencies in the use of municipal public resources and the delivery of public services in the area of urban mobility, energy, transport, ICT, and data management.[16]

As observed in D2.3, "Internal Report on Findings", the majority of business and management literature is focused on the firm, but not the city in its organisational form as the municipality. Whilst the business model could be considered as central to a theory of the firm, the same is not true of the municipality. Hence, the development of the REPLICATE city model canvas as a tool for municipalities to explore design, implementation, and governance issues, and so on to define the smart city business model as:

"...the way in which a city government organizes its services to create and deliver value for its citizens in a way that is economically viable, socially inclusive and environmentally sustainable".

"...while such smart services have the potential to deliver value to the city's residents, city governments cannot take for granted that they will in fact deliver value. This depends on how they are designed, implemented and governed; i.e. on the business models of smart services".

Assumptions still often take it that there is economic viability in what is being modelled. The question of what counts as value for citizens in the context of the delivery of services by a municipality is broad. Rodríguez Bolívar (2019)[17]

---

[16] "REPLICATE - D2.4 Report on the Replication Potential of City Business Models," REPLICATE Consortium, 2021, Accessed: Jul. 26, 2022. [Online]. Available: https://replicate-project.eu/wp-content/uploads/2021/12/REPLICATE_D2.4_Report_on_replication_potential_of_city_business_models.pdf

[17] M. P. Rodríguez Bolívar, "Public value, governance models and co-creation in smart cities," Public Administration and Information Technology, vol. 35, pp. 271–280, 2019, doi: 10.1007/978-3-319-98953-2_11.

argues that: "public value creation is the new lens for analysing smart cities. Based on participative governance models, local governments in smart cities must provide the tools and context to foster citizen engagement in public decisions and co-creation of public services. All this have to be addressed to increase the quality of life of citizens in their urban life".

Whilst the city model canvas developed in D2.2, "Report on the Business Models of the Lighthouse Cities"[10], and as transferred to the pilot cities, has provided a means for capturing a triple bottom line view of the city business model, there, nonetheless, remains the problem of connecting this perspective to the essentially financial view of businesses within industries expecting to supply solutions over the next 20–30 years. Businesses' single financial bottom line is the only one that eventually matters when it comes to maintaining a sustainable and viable firm. The industrial view presented from D9.3, "Sectorial Business analysis/Exploitation potential in the field of energy, ICT, sustainable mobility and other remaining sectors included in REPLICATE", certainly underlines this point in terms of the expected market size for certain types of solution.

"Big data and analytics as strategies to generate public value in smart cities: Proposing an integrative framework".[18]

The conceptualisation of business models in value creation ecosystems and the consideration of the municipality as the central actor in the "smart-city-as-a-network" [19](Yearworth, 2020) aligns well with the influential analysis approach of Amit and Zott (2001),[20] which positions business models as configurations of activities within strategic networks. In terms of our initial models of coordination, we are looking for hierarchical coordination to play a leading role, supported by networking and the market just where it can play its "normal" role.

The conclusions of the study have shown that achieving replicable business models is not the core problem to be solved in order to offer municipalities a way forward in their commitments to achieving climate change targets, arguing for a change in focus to the question of financing the capital

---

[18] F. Cronemberger and J. R. Gil-Garcia, "Big Data and Analytics as Strategies to Generate Public Value in Smart Cities: Proposing an Integrative Framework," Springer, 2019, pp. 247–267. Accessed: Jul. 26, 2022. [Online]. Available: https://link.springer.com/chapter/10.1007/978-3-319-98953-2_10

[19] Yearworth, M. REPLICATE - D2.3 Internal Report on Findings REPLICATE Project Deliverable (pp. 89), University of Exeter, 2020.

[20] R. Amit and C. Zott, "Value creation in E-business," Strategic Management Journal, vol. 22, no. 6–7, pp. 493–520, Jun. 2001, doi: 10.1002/SMJ.187.

investments required and the range of financing models that municipalities can implement to raise the necessary finance.

"The problem facing cities in scaling-up interventions to meet their climate change targets is not the absence of available capital in the world... Nor is there an absence of industries ready to sell technical solutions". "The problem is that to place the burden on municipalities to solve the business model problem would seem to be an abrogation of responsibility on the part of Governments, who should really be providing the necessary capital investment at city level to bridge the financing gaps that municipalities are facing".

11.10 RUGGEDISED Project

Previous sections have touched upon the process of value creation within business models. RUGGEDISED has derived a framework for understanding value creation through urban data platforms (UDPs). The proposed framework provides a systematic and comprehensive approach for understanding UDP adoption, use, and value creation by identifying key dimensions of UDPs and describing their effects on value creation.

This follows, on the one hand, from urban data platforms being regarded as a specific type of digital platform and key dimensions that determine the success of digital platforms are likely to also be relevant to a UDP. On the other hand, UDPs are part of smart city initiatives, representing the keystone that connects the digital technologies infused into city systems to handle growing urbanisation and keep cities liveable and thriving. Accordingly, the key dimensions of a smart city are also likely to be relevant to the success of a UDP. And as we have argued in this work, the addition of personal data into the equation can be regarded as the "icing on the cake".

RUGGEDISED undertook two literature reviews to cover these topics and provided insights concerning the key UDP dimensions. These all directly or indirectly relate to the business model required as they address the overall purpose for the UDP.[21]

1. What should be the purpose of an urban data platform?
2. Who should be the owner and manager of an urban data platform?

---

[21] S. Bagheri, T. Brandt, H. Sheombar, and M. van Oosterhout, "Value Creation through Urban Data Platforms: A Conceptual Framework," in Proceedings of the 54th Hawaii International Conference on System Sciences, 2021, pp. 2464–2473. Accessed: Jul. 26, 2022. [Online]. Available: https://aisel.aisnet.org/cgi/viewcontent.cgi?article=1576&context=hicss-54

3. What kind of value should be created with urban data platforms?
4. Why should urban data platforms be interoperable?
5. Why is citizen engagement needed for urban data platforms?
6. How can trust be fostered and who needs to be trusted?

(This work carried out by RUGGEDISED essentially laid the foundations for this book.)

Further, they focused upon another key dimension that has received considerable attention which is the platform business model.

"An important element of the institutional arrangement for an UDP is the funding of the platform. The availability of public funding and private financing resources as well as the return on investments for organizations investing in UDPs are crucial for the success of a UDP in a city ecosystem,[11,22,23] to create and contribute to economic, environmental, and social value– often referred to as the triple bottom line[12] – in a city ecosystem. Accordingly, the returns of investments of UDP projects are not always captured in monetary value and can be broadened".

Interoperability refers to the ability of different systems to interact and share information.[23] Lack of interoperability might lead to incompatible data exchange formats and protocols, which may in turn reduce the economies of scale of value added services using shared data of the platform.[24] A higher degree of interoperability also makes the UDP more appealing for new partners to join. Furthermore, UDP interoperability also facilitates data sharing between cities. We have addressed this issue further in Chapter 7.

A business model in these conditions needs to relate governance models with the mechanisms for generating income from the services it provides and how this income is utilised or shared (see also Chapter 14). And the sources for this revenue generation are countless. Examples from deployment of a smart card in a city environment showed that costs for deployment could be

[22] G. F. Camboim, P. A. Zawislak, and N. A. Pufal, "Driving elements to make cities smarter: Evidences from European projects," Technological Forecasting and Social Change, vol. 142, pp. 154–167, May 2019, doi: 10.1016/J.TECHFORE.2018.09.014.

[23] P. Pierce and B. Andersson, "Challenges with smart cities initiatives – A municipal decision makers' perspective," Jul. 2017. doi: 10.24251/HICSS.2017.339.

[24] M. de Reuver and W. Keijzer-Broers, "Trade-offs in designing ICT platforms for independent living services," 2015 IEEE International Conference on Engineering, Technology and Innovation/ International Technology Management Conference, ICE/ITMC 2015, Mar. 2016, doi: 10.1109/ICE.2015.7438645.

raised from unexpected sources – lower insurance premiums for a university using it for access control, reduced staff needed in a bus company due to faster turnaround of buses with quicker payment mechanisms, etc. Similarly, in deploying an UDP, pleasant surprises may be around every corner.

The identified sources of revenue by RUGGEDISED for the data platform focussed on[25]:

- payments by data providers or users in the form of a subscription – over a period of time;
- commission for each transaction;
- advertising.

A lesson to be learned when dealing with attracting personal data to a platform is the requirement to reach a critical mass. Final revenues for a platform are equally dependent on a successful recruitment strategy which could entail cross-subsidising, with a pricing structure to support this.[24,26,27]

It is expected that the establishment of the right pricing structure by the platform manager influences the adoption decisions of platform users and supports network effects within a platform ecosystem.[28] In relation to the models set out, here the manager is acting hierarchically, being in a position to make and implement such pricing decisions, which are transmitted through the pricing mechanism of the market. But equally, with a role of creating a growing eco-system, it is also likely that networking will be extremely important and decisions may be negotiated to achieve these aims. Thus, the "platform manager implements two types of control mechanisms: formal control (gatekeeping and process control) and informal control (e.g., shared norms and values) over the city stakeholders involved in the provisioning and utilization of urban data".

---

[25] R. Schüritz, S. Seebacher, and R. Dorner, "Capturing value from data: Revenue models for data-driven services," Proceedings of the Annual Hawaii International Conference on System Sciences, vol. 2017-January, pp. 5348–5357, 2017, doi: 10.24251/HICSS.2017.648.

[26] M. Schreieck, A. Hein, M. Wiesche, and H. Krcmar, "The challenge of governing digital platform ecosystems," Digital Marketplaces Unleashed, pp. 527–538, Sep. 2017, doi: 10.1007/978-3-662-49275-8_47/COVER.

[27] M. de Reuver, B. Nederstigt, and M. Janssen, "Launch strategies for multi-sided data analytics platforms," Proceedings of 26th European Conference on Information Systems, ECIS, 2018. Accessed: Jul. 26, 2022. [Online]. Available: https://pure.tudelft.nl/ws/portalfiles/portal/53104707/1360_doc_2.pdf

[28] M. Engert, M. Pfaff, and H. Krcmar, "Adoption of Software Platforms: Reviewing Influencing Factors and Outlining Future Research," Jul. 2019.

Informal control refers to the degree to which the platform manager relies on norms and values that it shares with all stakeholders involved in a platform ecosystem, and a similar set of values, beliefs, and shared norms provides a common foundation for the stakeholders within a city ecosystem. It can promote their commitments to the objectives of the UDP and encourage desirable behaviours.

11.11 Safe-DEED

The work carried out in the Safe-DEED project[29] provides a firm foundation for taking the next step in ascertaining how adding personal data into the mix can be valued and the topic will be returned to in Chapter 13.

Underpinning the project was the recognition of the weakness that "many companies have no data valuation process in place, Safe-DEED provides a set of tools to facilitate the assessment of data value, thus incentivising data owners to make use of the scalable cryptographic protocols developed in Safe-DEED to create value for their companies and their clients".

In terms of business model innovation, the objectives of Safe-DEED included quantifying and demonstrating the economic value for users and buyers of the developed privacy preserving technologies. The project developed "new multi-actor business models for privacy enhancing and data valuation technologies as well as a decision-support tool for designing and testing business models".

They explored the space spanned between existing threats to the confidentiality and privacy of user data and the actual data assets that users wish to see protected. By following a user-centric approach, they sought to ensure that the asset analysis was "not tainted with presuppositions of underlying needs of organizations, and instead focus on a utilitarian perspective contributing privacy enhancements for the good of end-users".

Before adding further actors, companies and third parties to the model "explore the practical incentive situation to identify a model that facilitates an ideal outcome for all involved parties".

They validated the model in two operational use-cases, (manufacturing and telecom) and open data, each with its own business model, making it

[29] "SAFE-DEED ." https://safe-deed.eu/ (accessed Jul. 22, 2022).

clear "how data owners can enhance the value they create and capture thanks to the technologies developed".[30,31]

These case-specific business models were then developed into generic business model designs, which subsequently formed the basis for a tool made available for free to entrepreneurs.

Another key aspect of Safe-DEED's work was related to private and public data. A work-package focused on "on the design and implementation of the Big Data Valuation component to be used to predict knowledge value of a certain corpus of a structured data without, however, having to completely analyse it". A task here was to develop "context-unaware valuation and context-aware valuation algorithms that will go into making the valuation tool and Implement the building blocks of the Big Data Valuation component".[32]

The Safe-DEED deliverable D4.3 report on the context-aware and context-unaware valuation[33] has an extensive review of the literature on the topic of data valuation methods. It starts from a tentative definition of data value around several key areas: contexts, data quality, privacy, aggregation, and reporting. It also discusses the properties that make data difficult to assess and brings valuable examples from data valuation applied to personal data. Focusing on the central notion of data quality, the document reviews a number of data quality assessment methodologies, discussing the diversity of data quality dimensions that they employ and the metrics that support their operationalisation. The report concludes with a discussion on the challenges of aggregating these aspects under a composite measure, and how reporting through certification or impact-based narratives can be a feasible alternative. We will return to this topic in Chapters 13 and 17.

11.12 The Safe-DEED Tools

- Safe-DEED data-driven business canvas[34]: The Safe-DEED data-driven business canvas is designed to help develop a data-driven service

[30] M. de Reuver, W. Agahari, R. Dolci, G. Breitfuss, and M. Fruewirth, "Safe-DEED - D2.2 Business models for use cases and generic business models," Safe-DEED Consortium, 2020, Accessed: Jul. 26, 2022. [Online]. Available: www.safe deed.eu

[31] G. Breitfuss, M. Fruhwirth, L. Disch, M. de Reuver, and W. Agahari, "Safe-DEED - D2.3 Business model decision support tool," Safe-DEED Consortium, Accessed: Jul. 26, 2022. [Online]. Available: www.safe-deed.eu

[32] "Data Value - Safe-DEED." https://safe-deed.eu/wp-data-value/ (accessed Jul. 26, 2022).

[33] M. Tufiş, "Safe-DEED - D4.3 Report on context-aware and context-unaware valuation Status Final," 2020, Accessed: Jul. 25, 2022. [Online]. Available: www.safe-deed.eu

[34] "Safe-DEED Data-Driven Business Canvas - Business Makeover." https://businessmakeover.eu/tools/safe-deed-data-driven-business-canvas (accessed Jul. 26, 2022).

innovation. The five main sections of the canvas support the structuring and concretisation of a data-driven use-case idea including first financial considerations. The provided descriptions and examples facilitate the work in interdisciplinary teams on future data-driven services.

- Safe-DEED data map[35]: The Safe-DEED data map helps with the identification of possible data sources that can be utilised to develop new data-driven services. It is divided into four quadrants. The four quadrants vary mainly in terms of ownership and usage rights of the data.
- Safe-DEED data service cards[36]: The Safe-DEED data service cards help enhance or develop new data-based services through the systematic combination of data sources, analysis methods, customer benefits and revenue opportunities. There are 50 cards as inspiration in the development process of data-driven innovations.

11.13 DUET Project

The DUET project focused on the creation of digital twins, and their contribution to business models focuses on the exploitation of these digital twins. Cities across Europe are using DUET to create their own digital twins for systemic policy impact exploration and experimentation based on policy-ready-data-as-a-service (PRDaaS) in a virtual environment. This advances global standards for city data enabling a digital twin to be set up in one-click for systemic policy impact exploration and experimentation.[37]

Essentially, the business model focuses on how DUET can improve the workings of a public administration, and the perceived benefits of a public sector twin include:

- improved operations;
- making service decisions based on a unified view of city processes;
- real-time information and predictive impact simulations;
- increased value for money;
- ability to experiment with different policy options online;

[35] “Safe-DEED Data Map - Business Makeover.” https://businessmakeover.eu/tools/safe-deed-data-map (accessed Jul. 26, 2022).

[36] “Safe-DEED Data Service Cards - Business Makeover.” https://businessmakeover.eu/tools/safe-deed-data-service-cards (accessed Jul. 26, 2022).

[37] “DUET - D7.6 Business and Exploitation Scenarios v1,” DUET Consortium, 2021, Accessed: Jul. 26, 2022. [Online]. Available: www.digitalurbantwins.eu

- exploring impacts across multiple domains, rather than running multiple costly on-the-ground pilots;
- better citizen experience;
- ability to act quickly and seamlessly on citizen real-time needs;
- help people understand why certain planning or operational decisions have been made.

11.14 DataVaults Project

The business model for the DataVaults project can be viewed from a variety of viewpoints:

- from that of a large company incorporating evolving technology and knowledge into its normal day-to-day activities;
- from the viewpoint of SMEs having a specific niche in their activities;
- how the project as a whole may collectively move forward on these views;
- how taking a "smart city exploitation approach" differs.

An example based on the conventional state of thinking from the DataVaults project at an early stage of its work was designed around having a product to exploit, rather than also being seen as an integral part of a smart city's evolution.

Two broad pathways towards commercial exploitation of the DataVaults outputs were explored:

- bringing into the market the DataVaults offering one or more personal data marketplace(s) backed up and operated by an entity that is composed of the various consortium partners, either in the form of a joint venture or individually;
- licensing the code and the DataVaults framework to interested organisations that would like to run the platform on their own and, in addition, providing consulting, training, and customisation services.

The early thinking around the business model was based on the lean canvas business innovation instrument, proposed by Ash Maurya.[38] This was the basis for creating a specific DataVaults' lean canvas, described below

[38] A. Maurya, Running Lean: Iterate from Plan A to a Plan That Works (2nd Ed.). O'Reilly Media, Inc., 2012, ISBN: 9781449305178.

The lean canvas is a simple one-page business plan template, similar to that of the business model canvas, and aims to deconstruct a business idea into its key assumptions. When compared with the business model canvas, developed by Alex Osterwalder, many practitioners agree that the lean canvas is more appropriate for start-ups or for bringing innovations to the market, as opposed to the business model canvas that functions better as a descriptive canvas for an existing business model, with the scope of sharing knowledge with the whole team within an enterprise.[39]

The sectors to be completed are: customer segments, the problem to be solved, existing alternatives, the unique value proposition, high level concept, solutions, channels, revenue streams, early adopters, cost structures, key metrics, and "unfair advantage held".

The high-level concept part of the canvas aims to provide a high-level description of the value proposition for users, often drawing a parallel to a well-known service/product that the customers know quite well. For example, DataVaults, as a high-level concept, can be seen as an "eBay" for personal data from the side of data owners, where they can decide what to sell and how much, keeping also part of their anonymity. On the other side, data seekers can think of DataVaults as a market research service that is directly targeting and pulling data from individuals, once they consent to share these data.

Key metrics identified included:

- number of available datasets;
- the mean value of compensation per data transaction paid out;
- the size of the data owners population;
- the size of the population of the data seekers.

11.15 Viewpoint from a DataVaults SME's Perspective

11.15.1 Assentian

Returning to our models, a simple example of how just the market might be used for a typical SME engaged in the trading of personal data can be seen in this analysis of the business opportunities and basis for the business model.

In regard to the opportunity, the technology on its own has limited value. What the technology does is enabling access to where the derived

[39] I. Jeffries, "How To Fill In A Lean Canvas ," 2020. Accessed: Jul. 26, 2022. [Online]. Available: https://isaacjeffries.com/blog/2020/9/23/how-to-fill-in-a-lean-canvas

value resides. We provide a capability for data owners/custodians to share and monetise data. The attractiveness lies in:

- regulatory compliance;
- maintaining control;
- transparency over who sees what;
- access to big pools of data which can support commercial planning, service delivery, research and development, etc.

But the SME is faced with determining the costs of providing such a service, of making the data accessible in order to share it. It is hard to provide concrete figures as data comes from diverse sources and is stored in different ways and forms. Cost estimates have to be made based on some general characteristics of data source/type, examples of which are:

- mobile app data;
- social media data;
- GIS data;
- healthcare data (from public/private healthcare systems);
- public sector data (held by local authorities and/or central government departments in data centres/cloud and on proprietary systems typically), etc.

Clearly, the cost of access to enable sharing would have to be lower than the perceived monetary value or benefit. This value can be derived through a number of potential criteria:

- utility of data;
- business value and/or economic value;
- performance value;
- urgency;
- validity;
- scarcity;
- coverage;
- useful life;
- market value – what is the willingness to pay and the demand for it?

The range of revenue models under consideration include:

- access to data sharers for free;
- the percentage of revenue earned by sharers is fee charged to facilitate sharing;
- regular data sharers pay a monthly fee for access and additional cost(s) for high volume utilisation;
- 100% of sales revenue kept by data sharers.

The data buyers can be categorised as either being "one off users", where a single fee is paid to access the platform, and those regular users who pay a monthly or annual fee.

11.15.2 Andaman7

Andaman7 operates in the health sector and their approach is summed up as:

"The solution touches on a universal problem: human health and our Business Model is based on a wide distribution of the application in "Freemium". As such, revenues come from paid advertising and add-ons (including clinical studies apps). Our application is currently well differentiated and complementary to existing solutions. We do not want to be profiled as a competing solution to EHR."

Freemium is a business model in which a company offers basic or limited features to users at no cost and then charges a premium for supplemental or advanced features.

11.16 Digital Twins and Business Models

A study into business models and implementation challenges facing a digital twin is "Urban digital twin: Business models and implementation challenges".[40]

Urban digital twins are a virtual representation of a city environment with bi-directional communication links. They require collaborations between different actors in the urban ecosystems in order to provide a complete picture of

---

[40] R. D'Hauwers, N. Walravens, and P. Ballon, "From an Inside-In Towards an Outside-Out Urban Digital Twin: Business Models and Implementation Challenges," ISPRS Annals of the Photogrammetry, Remote Sensing and Spatial Information Sciences, vol. VIII-4/W1-2021, pp. 25–32, 2021, doi: 10.5194/isprs-annals-VIII-4-W1-2021-25-2021.

the situation in the city. In order to define the complex relationships between the different actors in the urban digital twin ecosystem, the business model literature helps to answer questions on how value can be created, and how the value network can be controlled. In this chapter, we identified four different types of business models for urban digital twins based on whether they are used by the government or the ecosystem, and whether the government or the ecosystem controls the value network of the urban digital twin. Interviews were held in five different existing urban digital twins to identify which challenges the different existing digital twins have when implementing the urban digital twins.

The outcomes of the business model scenarios support the design of urban digital twins:

- by identifying which decisions need to be made by cities when developing urban digital twins;
- by proposing cloud requirements for technology providers supporting cities, in the development of urban digital twins.

11.17 Conclusion

In this chapter, we have witnessed a wide range of approaches, all working in the same direction and contributing to the discussion of how best to take forward data platforms within a smart city and how these will make the most of scarce resources and we will return to this in subsequent chapters.

What is significant is the amount of information which collectively these projects featured above have produced on this topic and derived from hundreds of sources in their research. Many provide a range of tools and advice far in what you would expect to find on a project website. For example, DUET[41] provides a whole range of useful information divided into topics such as policy-making, technology, etc.

Most other projects will go into similar effort to make their findings readily accessible and hence useful to practitioners in cities across Europe.

[41] "Results - DUET Digital Urban European Twins." https://www.digitalurbantwins.com/outputs (accessed Jul. 26, 2022).

12

(Digital) City Financing Platforms

Juergen Ritzek

EEIP, Brussels, Belgium
Email: juergen.ritzek@ee-ip.org

Abstract

This chapter introduces the theme of City Financing Platforms, particularly digital ones, as a new and important tool to support the development of smart cities, mainly small- and medium-sized ones which might have several problems in finding the necessary funding for their innovative projects.

First, the role of such platforms is described, particularly as far as energy service companies (ESCOs) and project developers are concerned, who might benefit from such platforms to borrow money in the initial stage of their project and to find co-investors to support their activities, but also to get assistance by expert teams to make their projects financially attractive.

Second, some examples of existing financing platforms are described, according to the type of financial services they provide. Some key points on challenges and failings are also introduced in the conclusions, in particular the need to support cities at local level which are developing smart city strategies and plans, as emphasised by many regional, national, and EU-wide programmes.

12.1 Introduction

(Digital) City Financing Platforms represent an alternative route of financing city projects especially for small- and medium-sized cities with limited lending capacity. However, while the usage of such platforms requires the integration of the platform requirements early in the development phase of city projects, the platform market is itself in an early stage.

The following represents an outcome of the work till today of the City Financing Platform initiative as part of the Smart City Market Place, based

DOI: 10.1201/9781003399384-13

on online exchange amongst initiative members as well as webinars, conferences, and round tables.

12.2 Role of Financing Platforms

Financing is often stated as one of the biggest challenges to implement a smart city strategy. This is not only due to pressure on public budgets but also to inherent complexity of smart city projects often covering a range of different areas at the same time, e.g., energy efficiency measures for various types of building and the development of smart grids and integrated mobility concepts. Bigger cities might be able to develop cross-cutting robust business models, creative revenue models, and identify value capture which can attract investors. But this route is barely impossible for small- and medium-sized cities purely due to a lack of organisational capacity.

This is where digital financing platforms can play an important role, especially as the project developer or an energy service company (ESCO) – and not the city itself – would usually be the contractual partner of a digital financing platform.

Digital finance is increasingly demonstrating its ability to overcome key barriers for smaller scale project finance through the ability to make the processing and analysis of larger amounts of data cheaper, faster, and more accurate. This reduces the risk of obtaining timely, material information relevant to sustainability impacts and financial risks related to the investment while increasing transparency. However, platforms offering financing for smart city projects are still at a very early stage.

In principle, project developers and ESCOs can utilise such platforms for various purposes, e.g., to:

- borrow money to cover initial costs of project development;
- find co-investors that would take equity in the project;
- sell shares of the project at some stages of the activity.

The platforms themselves usually cover all or several of the following steps:

- deal sourcing and matchmaking;
- expert and knowledge management through standardised due diligence tools;
- project bundling for investors.

12.3 But Who are These Digital Financing Platforms – Or Where are They?

That is the big question. Very few seem to exist today, but a growing number is in development driven by the need to mobilise private capital to support sustainable growth.

Over the time, it became clearer that there is no easy solution available. What was clear from the very beginning is that such solutions would start becoming available only, or at least first, for certain kind of projects, namely for re-financing energy efficiency/deep retrofitting and renewable energy projects delivering a predictable revenue stream from energy savings or produced energy. And this revenue streams are in principle suited to allow re-financing of even upfront project financing.

12.3.1 Examples of digital financing platforms

Digital financing platforms usually provide one of the two financial services:

1. loans or credit lines – to finance the purchase of goods or services (loans) or to cover delays between receipts and payments, and to deal with specific periods of lack of liquidity or for specific purchases (credits);
2. re-financing based on calculated or actual revenue streams from projects.

12.3.2 Credit/loans

The most prominent example for this service is likely the Jack Ma's 3-minute loan, a service through which 16 million SMEs in China have received 300 billion USD in loans. And there are already some fintech companies providing a similar service with loans up to 300.000 USD provided to SMEs within a few hours purely using an online service. Examples are as follows:

- Ondeck, US:[1] loan service for SME, access to finance within a day.
- Forward Financing, US:[2] working capital for SME, access to finance within hours, up to USD 300k.
- Kabbage,[3] US: credit lines up to USD 150k within minutes (own claim). Owner: American Express.

---

[1] https://www.ondeck.com/company/

[2] https://www.forwardfinancing.com/

[3] https://www.kabbage.com/

12.3.3 Re-financing

Businesses operating digital re-financing platforms usually do not provide financing themselves but support the bankability of projects and/or help in getting access to finance. Although they cover very different types of projects and technologies, as of today, none of them has already become a mass market solution, mainly due to the complexity inherent in most projects seeking re-financing.

The list is clearly not complete but represents a good overview of current re-financing platforms:

- TrustEE:[4] Likely the most advanced platforms with software supported project evaluation features up to a linked special purpose vehicle (SPV) directly providing finance. Focus is more on industrial projects but in principle also applicable for smart cities projects.
- EEnvest:[5] Risk evaluation tool for energy efficiency investments in buildings.
- eQuad:[6] Bridging platform for projects and investors.
- ESI:[7] Project structure modelling platform including insurance.
- SOURCE:[8] Software-enabled project structuring and implementation platform, financed by regional development banks.
- Optimised Retrofit: Project delivering a procurement framework but also trainings with a focus on buildings.

Next to platforms using a standard technology approach, first, start-ups are on their way to use blockchain technology for tracing and tokenising energy savings. Well, there is a lot of hype around blockchain and it is not always easy to distinguish between what really works (or can work) and what is just a nice white paper. The two most prominent attempts are as follows.

- EFFORCE:[9] A lot of big buzz including a celebrity as a door opener, Apple co-founder Steve Wozniak. In principle, EFFORCE covers

[4] https://trustee.ise.fraunhofer.de/about
[5] http://www.eenvest.eu/
[6] https://www.eu.jouleassets.com/whitepaperbridgingthegap
[7] https://www.esi-europe.org/financing/
[8] https://regions20.org/source-online-software-structuring-projects/
[9] https://efforce.io/

everything. After the implementation of an energy efficiency project, energy consumption is tracked with smart meters delivering data to a smart contract for initiating payments. Projects get bundled and tokenised and everybody can buy such tokens (cryptocurrency WOZX) and benefit from regular returns. Tokens can also be traded on some crypto exchanges providing a secondary market and finally liquidity.

- Energyglare:[10] Similar approach to EFFORCE. Energyglare plans to build a platform (second half of 2022) with a third-party verifier level for planned energy savings by implementation service providers, tools for institutional investors to build portfolios matching capital deployment requirements, a reporting layer, as well as an interface to convert energy savings into tokenised carbon credits (2024).

12.3.4 Challenge project pipeline: the chicken and egg problem

The lack of projects represents the key challenge for digital city financing platforms – and not the availability of private funds.

Financing platforms can deliver an easy access to private finance including the management of the transactions following project implementation. But they rely on two factors they can hardly influence:

- the availability of bankable projects;
- project structuring fitting the requirements of financing platforms/ investor.

Both are deeply rooted in city planning processes and a lot of information about smart city project challenges, solutions guidelines, and reports can be found on the EU Smart City Marketplace website. Good practice solutions have been already identified many years ago such as the case of Cerdanyola (2013), where experts from universities and consulting companies were involved in preparing feasibility studies and providing scientific methods for the set of requirements included in the call for tender for the ESCO.[11]

Many regional, national, and EU-wide programmes, including the next Smart Cities Marketplace period 2022–2025, are putting specific emphasis

[10] https://energyglare.io/

[11] https://smart-cities-marketplace.ec.europa.eu/insights/publications/executive-summary-report-planning-and-implementation-process-assessment, p137 and p144–145, accessed April 2022

on supporting cities at local level which are developing smart city strategies and plans towards individual project planning and tendering procedures.

Projects represent the "fuel" for city financing platforms, as they are designed to deliver private finance to many projects while reducing financing costs, time to set-up financing, and the need for public funds, all of which is a pre-requisite to scale up smart city project implementation, urgently required to tackle climate change, energy security, and affordability.

12.4 Conclusion

There is no "silver bullet".

Clearly, financing cannot be seen in isolation and a 3-minute financing solution for smart city projects might also not be available overnight if ever, even if the smart city project pipeline would be filled.

On the other hand, various platforms have started to develop comprehensive solutions for all engaged stakeholders instead of only focusing on a single challenge, e.g., including green taxonomy and reporting layers, investor portfolio management, or third-party verification and insurance solutions.

In addition, blockchain technology can help driving costs further down, offering several automation options such as automatic transfer of data via smart meters to smart contracts initiating payments, but there are still many open questions such as data security and standards on oracle/smart contract level. Tokenisation is a key feature of blockchain/DLT and can boost liquidity though creation of secondary markets, but the use of cryptocurrencies is still in its infancy.

Digital city financing platforms are, although still in their early phase, not a bottleneck for smart city project finance. Their business model is easy and market driven focusing on capturing a share of the investment or project return – what the business model depends on is the availability of many bankable projects, currently not available.

The bottleneck today is the project pipeline. In other words, money is not a problem.

13

The Governance of Personal Data for the Public Interest: Research Insights and Recommendations

Marina Micheli

Scientific Project Officer, Digital Economy Unit, European Commission's Joint Research Centre (JRC), Ispra, Italy

Abstract

The chapter describes the emerging approaches for the governance of personal data with a focus on the role city administrations might play in promoting a more inclusive data landscape. It illustrates the findings of social science research conducted at the Digital Economy Unit of the Joint Research Centre of the European Commission. It is composed of an introduction and three sections. After describing four models for the governance of personal data, the chapter presents an empirical research based on interviews with cities' chief data officers, then concludes with key recommendations. The findings presented in the chapter show that city governments could play a key role in addressing power unbalances of the current data landscape, acting as trusted data intermediaries and enabling the use of citizens' personal data for the public interest. To conclude, the chapter describes six organisational strategies that city practitioners can adopt to enhance personal data sharing for the public interest through inclusive approaches.

13.1 Introduction

The chapter discusses emerging approaches for the governance of personal data, with a focus on the role city administrations might play in promoting a more inclusive data landscape. Peoples' digital footprints, generated as a by-product of their daily activities and collected by private and public organisations, are increasingly extensive. However, the opportunities that citizens

DOI: 10.1201/9781003399384-14

have to control how these are used and for which purposes is still limited. A more inclusive approach to data governance would allow citizens, as well as other actors, to have a greater say in how data is used and foster socially relevant usages of their data.

In this chapter, drawing from recent work conducted at the *Joint Research Centre (JRC) of the European Commission*, I sketch some of the key issues at stake in relation to the governance of citizens' data and provide recommendations based on lessons learned from initiatives led by the Digital Economy Unit of the JRC. The chapter is structured as follows:

- The first section presents emerging conceptualisations, prototypes, and practices of <u>models for the governance of personal data</u>. These models offer ways for accessing, controlling, sharing, using, and deciding about data that are "alternative" to the dominant approach – promoted especially by Big Tech platforms – allowing greater manoeuvre to citizens, civic society, and public sector organisations.
- The second section discusses a particular way in which city administrations can address asymmetries of the current data landscape: business to government (B2G) data sharing, which implies <u>public authorities accessing privately held data of public interest</u>. This data relation represents a paradigm shift in the understanding of data flows of the public authorities. Traditionally understood as a source of information *for* the private sector (with open data), now public authorities are acting (also) as recipients of data flows *from* the private sector (with B2G data sharing, see also the notion of reverse public sector information).[1]
- The last section includes a few lessons learned and recommendations. These are not meant to be exhaustive, but to provide some ideas and practical guidance (based on the aforementioned research) to local pubic administrations engaging in data innovation projects with a citizen-centric perspective.

---

[1] Poullet, Y. (2020). From open data to reverse PSI: a new European policy facing GDPR. European Public Mosaic, (11), 42–57.

13.2 Alternative Models for Data Governance

In the context of the project Digitranscope,[2] conducted at the Centre for Advanced Studies[3] of the JRC, we explored the emerging approaches for personal data sharing, control, and use put forward by a wide range of stakeholders in Europe. We were interested in data governance models imagined or implemented across Europe that offer an alternative to the dominant "take it or leave it" approach for handling personal data typical of big tech platforms (Craglia *et al.*, 2021).[4] Instead of addressing issues related to risks and protection of personal data, we explored the opportunities for agency and increased control of data.

We conducted a review of relevant resources from the scientific and grey literature and came up with four emerging models for the governance of personal data, which we labelled: data sharing pools, data cooperatives, public data trusts, and personal data sovereignty(for a comprehensive overview see (Micheli *et al.*, 2020))[5].

In the article, we describe the models mainly in abstract terms emphasising the power social actors have to control how data is accessed and used to produce value. The models are heuristic tools, useful to understand and further examine the practical implementations of emerging approaches to data (see Figure 13.1). In the last few years, various organisations made similar attempts to systematise emerging practices of data governance; see, for instance, GovLab's data collaboratives explorer,[6] NESTA's new ecosystem of trusts (Mulgan & Straub, 2019),[7]

[2] Craglia, M., Scholten, H.J., Micheli, M., Hradec, J., Calzada, I., Luitjens, S., Ponti, M. and Boter, J., *Digitranscope: The governance of digitally-transformed society*, EUR 30590 EN, Publications Office of the European Union, Luxembourg, 2021, ISBN 978-92-76-30229-2 https://publications.jrc.ec.europa.eu/repository/handle/JRC123362

[3] https://joint-research-centre.ec.europa.eu/knowledge-research/centre-advanced-studies_en

[4] *As 2, above*

[5] Micheli, M., Ponti, M., Craglia, M., & Berti Suman, A. (2020). Emerging models of data governance in the age of datafication. Big Data & Society, 7(2), 2053951720948087. https://doi.org/10.1177/2053951720948087

[6] https://datacollaboratives.org/explorer.html

[7] Mulgan, G, Straub, V (2019). The new ecosystem of trust: How data trusts, collaboratives and coops can help govern data for the maximum public benefit. Nesta. Available at: https://www.nesta.org.uk/blog/new-ecosystem-trust/

Model	Key actors	Goals	Value	Mechanisms
Data sharing pools (DSPs)	• Business entities • Public bodies	• Fill knowledge gaps through data sharing • Innovate and develop new services	• Private profit • Economic growth	• Principle of 'data as a commodity' • Partnerships • Contracts (e.g. repeatable frameworks)
Data cooperatives (DCs)	• Civic organisations • Data subjects	• Rebalance power unbalances of the current data economy • Address societal challenges • Foster social justice and fairer conditions for value production	• Public interest • Scientific research • Empowered data subjects	• Principles from the cooperative movement • Data commons • 'Bottom-up' data trusts • GDPR Right to data portability
Public data trusts (PDTs)	• Public bodies	• Inform policy-making • Address societal challenges • Innovate • Adopt a responsible approach to data	• Public interest • More efficient public service delivery	• Principle of 'data as a public infrastructure' • Trust building initiatives • Trusted intermediaries • Enabling legal framework
Personal data sovereignty (PDS)	• Business entities • Data subjects	• Data subjects self-determination • Rebalance power unbalances of the current data economy • Develop new digital services • centred on users need	• Empowered data subjects • Economic growth • Private profit • Knowledge	• Principle of 'technological sovereignty' • Communities and movements (e.g. MyData) • Intermediary digital services (personal data spaces) • GDPR Right to data portability

Figure 13.1 Summary of emerging data governance models. Source: Micheli et al., 2020, p. 6

and Mozilla's "Database of Initiatives - Alternative Data Governance in Practice".[8]

Three of the four emerging models are based on the notion of trusted data intermediary. They aim to enhance individuals' control over their personal data collected by various actors (from platforms to governments), on how it is shared and used, to unlock its value for the single individual, communities, and society. Yet, the article also shows that they pursue different goals. In brief, these models are as follows.

- Personal data sovereignty based on a new kind of digital services (called personal information management systems, personal data spaces/stores, etc.), which are competing on the market and whose goal is to empower individual citizens in their ability to choose to what use put their data. They allow users to store, aggregate, and decide how to share data with third parties.
- Data cooperatives are grassroots-driven decentralised organisations in which members of certain communities voluntarily pool their data together. Based on a critique of the extractivist model of platform capitalism, they provide democratic control over data, allow members to voice their needs and concerns, and produce societal benefits.
- Public data trust is a model in which a public actor (such as a local public administration) establishes a relationship of trust with citizens and manages data on their behalf. A public sector entity assumes the role of trustee to guarantee that citizens' data is handled ethically and securely, while enabling the re-use of data for public interest purposes.

As the last model suggests, local public administrations could play a key role in addressing power unbalances of the current data landscape acting as trusted data intermediaries and enabling the use of citizens' personal data for the public interest. Public data trusts are more of a prototype than actual practices, as there are still limited experiences of this model. Yet, they offer an ideal from which city administration can get inspiration. Importantly, they imply the establishment of a relationship of trust between citizens and public bodies: citizens must be reassured that public actors are capable to keep their personal information safe and secure and that they will use it to improve their lives and society. To earn trust, public bodies might engage in citizens' consultations and living labs, launch initiatives for citizens digital

[8] https://foundation.mozilla.org/en/data-futures-lab/data-for-empowerment/who-is-innovating-database-of-initiatives/

rights (e.g., Wiltshire and Pierri, 2021),[9] ask the intervention of external independent organisations that act as trusted intermediaries, and disseminate best practices and achievements.

A pioneering initiative promoting cities as data trusts was the EU Horizon 2020 project DECODE, which was meant to return the value of personal data back to citizens while increasing their control over how data is shared (Bria and Morozov, 2018; Decode, 2019[10]). The project developed a privacy-protecting platform for citizens' participation and a cryptographic technology (based on distributed ledger technology) that allows citizens to control how to share sensors data. The city of Barcelona also pioneered the adoption of data sovereignty clauses in public procurement to establish the city's right and mandate to acquire the data generated through public contracts (Monge *et al.*, 2022),[11] which is the topic of the following section.

13.3 City Administrations' Access to Personal Data of Public Interest

Cities are a key actor for enabling more inclusive data governance, not only because they can support citizens' digital rights and data control but also because they can foster data (re)use for the public interest. Citizens constantly "leave" digital footprints as by-products of common everyday activities, which are used by private sector companies to collect a great amount of personal data potentially beneficial for city administrations to address health, societal, and environmental challenges. For instance, public authorities have recently sought access to location data collected by mobile phone operators to tackle the spread of the COVID-19 virus, monitor compliance to restrictions, and for urban traffic management. Yet, substantial asymmetries exist between big tech corporations and local governments for what concerns control and access to data of public interest. Public bodies, especially local and regional governments, are in a weaker position vis a vis tech companies and are struggling to find sustainable ways to access information of public interest collected by those.

[9] https://www.onlineopen.org/media/article/583/open_essay_2018_morozov_rethinking.pdf

[10] DECODE (2019). D5.9 Final report on the Barcelona Pilots, evaluations of Barcelona Now and sustainability plans. Available at https://decodeproject.eu/publications/final-report-barcelona-pilots-evaluations-barcelonanowand-sustainability-plans.html

[11] https://eprints.qut.edu.au/232522/

In the article "Public bodies' access to private sector data" (Micheli, 2022),[12] I examine the perspectives and experiences of managers and project leaders of 12 European municipalities on access to private sector data of public interest. Drawing from interviews, I discuss the most common ways through which local public administrations have access to privately held data and what their advantages and drawbacks are according to those working in the field. Administrations adopt different models, but most of their initiatives are pilots' projects and access to private sector data is still a niche experience due to the lack of incentives and technical, organisational, and economic obstacles (Martens and Duch-Brown, 2020).[13]

In the article (Micheli, 2022), I describe four common approaches for data sharing[14] between business and cities, which are based on different kinds of relationships and lead to diverse outcomes for the administrations (Figure 13.2). The adopted approaches for business-to-government data sharing are as follows.

- **Data donorship (or data donation):** Private companies share data at no cost on a voluntary basis (often for corporate social responsibility). This occurs more often during emergencies or for humanitarian purposes.
- **Public procurement of data:** Local governments purchase data through *ad hoc* contracts with private companies, such as telecom operators, who allow them access to dedicated dashboards, reports, or data assets.
- **Data sharing pools:** Local governments establish "win–win partnerships" with private companies at no cost, based on the mutual sharing, and eventually joint analysis, of data.
- **Data-sharing obligations:** Local governments include "data sovereignty clauses" as part of subcontracted services specifying that data gathered by a service provider (e.g., public transport, waste management, and ride-sharing companies) is available and accessible, in a privacy-compliant way, to the city council.

[12] Micheli, M. (2022). Public bodies' access to private sector data: The perspectives of twelve European local administrations. First Monday, vol. 27, n. 2. https://firstmonday.org/ojs/index.php/fm/article/view/11720

[13] Martens Bertin and Duch-Brown Néstor, 2020. The economics of business-to-government data sharing. JRC Digital Economy Working Paper, 2020-04, at https://ec.europa.eu/jrc/sites/jrcsh/files/jrc119947.pdf, accessed 9 May 2021

[14] Other approaches are research partnerships with universities or research institutions and urban challenges or hackathons (HLEG, 2020).

	Role of local administration	**Enablers**	**Discourses**	**Outcomes**
Data donorship	Recipients	A city's reputation	"Incidental partnership" "Ethical dilemma"	Divide between cities
Public procurement of data	Clients	Willingness to pay and negotiate	"Evaluation phases" "Reluctance"	Limited engagement, lack of data quality, and granularity
Data partner-ships and pools	Partners and collaborators	Data culture Professional network Capacity Socially relevant projects	"Win-win collaborations" "Productive relations"	Control of data access and use for advanced cities
Data sharing obligations	Administrators	Data culture Renewal of contracts	"Systematic sovereignty" "Data sovereignty"	Control of data access for subcontracted city services

Source: Micheli, 2022.

Figure 13.2 Summary of the approaches adopted by local administrations to access private sector data of public interest.

The findings of the research point to a "divide" between municipalities, concerning their chances of finding private companies willing to engage in data sharing. Bigger cities and "smart cities" are in a favourable position to access private sector data of public interest. Not only do they have more resources, bigger networks, and greater experience to establish partnerships and data pools, but private companies are also more likely to let them access data for free (via data donorship) as this gives them higher reputation and visibility. Private companies, in fact, might adopt such "use cases" to market their services and products to other – less -well-known – cities.

City managers are in favour of setting up collaborative partnerships at no cost with the private sector, based on collaboration and "co-creation", instead of spending resources and acquiring data via public procurement, as this approach also leads to better outcomes. Yet, public administrations with more resources and networks are more likely to establish such win–win collaborations. *A more inclusive approach is that of data sharing obligations*, which are clauses that municipalities can include in their tender contracts for subcontracted services. The clauses demand that data collected by a company as a by-product of delivering a public service is made accessible to the municipality (Bass *et al.*, 2018).[15] Local administrations often have kept control of the modalities through which privately held data is shared, so as to achieve greater data quality and granularity.

Once access to data is achieved, however, the question of how data is used to serve citizens' needs remains. To what extent the (re)use of such data assets has an impact, and of what kind, on those from which the data comes from (and others)? Are the interests of vulnerable or less privileged groups protected and how? Overall, do the efforts necessary for accessing data and putting in place the required infrastructure (technical, legal, and operational) pay off in terms of outcomes and social benefits deriving from its use? These (and other) questions should be asked when big data are used to inform policy-making – for a more comprehensive overview of demands for computational social science for policy, cf., Bertoni *et al.*, 2022.[16]

---

[15] Theo Bass, Emma Sutherland and Tom Symons, 2018. Reclaiming the smart city. Personal data, trust and the new commons, Nesta (23 July), at https://www.nesta.org.uk/report/reclaiming-smart-city-personal-data-trust-and-new-commons/ , accessed 9 May 2021.

[16] Bertoni, E., Fontana, M., Gabrielli, L., Signorelli, S. and Vespe, M. (2022), Mapping the Demand Side of Computational Social Science for Policy, EUR 31017 EN, Publications Office of the European Union, Luxembourg, 2022, ISBN 978-92-76-49357-0 (online),978-92-76-49358-7 (print).

13.4 A Few Recommendations for Cities

In this section, insights deriving from policy discussions and scientific research on the governance of personal data for public interest are presented. Most are based on research (with social science lenses) conducted by or for the Digital Economy Unit of the Joint Research Centre (EU Commission), which includes a close reading of the wider debate on these topics. The points below are practical recommendations for experts and managers in local administrations who wish to enhance personal data sharing and use, for the public interest through fair and inclusive approaches.

1. **Hire a data steward:** To implement data innovation, cities need to develop internal capacity and resources should be available for recruiting new professional roles in the organisations. Building internal technical, administrative, legal, and strategic capabilities includes setting up new specific managerial roles and recognised functions, such as that of "data stewards". Data stewards – one of the recommendations of the high-level expert group on B2G data sharing – are "individuals or teams that are empowered to proactively initiate, facilitate and coordinate" data sharing (European Commission, 2020).[17] Their role would be to systematise data partnerships and scale efforts; hence, they will have the expertise for promoting data access, sharing, and management.
2. **Take into account organisational barriers:** Silos between public offices and departments often act as barriers for data innovation at the local level. The creation of multi-stakeholders local data ecosystems is hindered by lack of communication and collaboration within public sector's offices, for instance, between local public transportation and the city government. Coordination and collaborations within public sector organisations and public offices are thus a key prerequisite for any form of digital innovation. Yet, the relational, cultural, and organisational challenges in setting up data ecosystems are at times underestimated, compared to technological and legal issues (compliance to GDPR) – as resulted from a recent qualitative study on local data ecosystems in selected EU cities (Liva *et al.*, 2022).[18]

---

[17] European Commission, 2020. Towards a European strategy on business-to-government data sharing for the public interest. Final report prepared by the High-Level Expert Group on Business-to-Government Data Sharing, at https://op.europa.eu/en/publication-detail/-/publication/d96edc29-70fd-11eb-9ac9-01aa75ed71a1

[18] Liva Giovanni, Micheli Marina, Schade Sven, Kotsev Alexander, Gori Matteo, Codagnone Cristiano (2022). City data ecosystems between theory and practice: A qualitative exploratory study in seven European cities (forthcoming).

3. **Understand internal demand and needs:** The perspectives and experiences of employees who are directly involved and/or impacted by data innovation should be taken into account in organisations' data innovation plans. Social research shows that there might be diverging visions between managers and front-line workers (such as counsellors and caseworkers) on the implementation of new data practices, which can even lead to gaming the systems or strikes by the latter (Dencik, 2022).[19] According to research, at times, preconceptions and "top-down pressures" (innovation "pushed" by organisations) might not encompass the needs of public sector workers. Therefore, it is important to listen to all perspectives, including those that might be critical, to successfully implement new data practices. Different levels and offices of local governments might have diverse needs; these should be accounted for, to avoid clashes or conflicts.

4. **Try a sandbox approach:** Sandboxing implies creating a "safe environment" for experimenting new technical infrastructures, organisational approaches, and/or legal schemes in order to facilitate data sharing and innovation. Traditionally, sandbox refers to the process of quickly and safely developing and testing new technical applications before operationalising them. Yet, more recently, the term is adopted in different contexts, such as in the case of organisational and regulatory sandboxing. Organisational sandboxes are carried out to develop and test solutions that address non-technical matters, for instance, through experimenting best practices, incentives, and terms and conditions for business to government data sharing (Kotsev *et al.*, forthcoming).[20] These sandboxes are based on the analysis of local data ecosystems and stakeholders' roles and incentives, as well as on the involvement of key stakeholders from different domains to discuss, plan, and develop new organisational solutions. Similarly, regulatory sandboxes are spaces for the experimentation of innovative solutions in a (near) real-world environment (Council of the European

---

[19] Dencik L. (2022) "Understanding demand for data-driven innovation in the public sector – the case of algorithmic processes" in Granell C., et al, Emerging approaches for data-driven innovation in Europe. Publications Office of the European Union, Luxembourg, ISBN 978-92-76-46937-7, doi: 10.2760/630723, JRC127730, pp. 93–98.

[20] Kotsev, A., et al (2022). Sandboxing: what it is and how to use it to strengthen your data ecosystem. Publications Office of the European Union, Luxembourg, Publications Office of the European Union, Luxembourg (forthcoming).

Union, 2020).[21] Together with test beds and living labs, these experimentation spaces offer a setting in which different stakeholders (including citizens) can co-develop solutions and associated regulations, through relationship of trust and knowledge exchange (see Kert, Vebrova, & Schade, 2022).[22]

5. **Form or join alliances between cities:** Cities are increasingly "joining forces", and collaborating within networks, groups, and alliances, to enhance their opportunities to access, use, and better govern (personal) data for the public interest. For instance, from our interviews about B2G data sharing with practitioners with European local administrations, it emerges that, as a common strategy to enhance negotiating power for accessing private sector data, cities are building alliances and networks (Micheli, 2022). Although the vast majority of B2G data sharing still consists in bi-lateral relations (between a single municipality and a data holder company), practitioners share the belief that to make the process more efficient and fair, they have to work with other cities. Some respondents take part in national or transnational networks to address some of the challenges of B2G data sharing. For instance, a city's chief data officer collaborates with a national association of municipalities to develop a joint standard contractual framework for data sharing relations with private companies. In Europe, there are various formal associations of cities that are promoting citizen-centric approaches to data governance (of which the reader of this report is surely familiar with). These include, for instance, Eurocities,[23] the Living-in.EU,[24] the Council of European Municipalities and Regions,[25] and the Cities Coalition for Digital Rights.[26] Several European projects are also meant to bring cities together, enabling them to collaborate, innovate, and

[21] Council of the European Union (2020). Regulatory sandboxes and experimentation clauses as tools for better regulation: Council adopts conclusions. Press release, 16 November. Available at: https://www.consilium.europa.eu/en/press/press-releases/2020/11/16/regulatory-sandboxes-and-experimentation-clauses-as-tools-for-better-regulation-council-adopts-conclusions/

[22] Kert, K., Vebrova M. & Schade, S. (2022) Experimentation spaces for regulatory learning: Test beds, living labs, and regulatory sandboxes. Science for policy brief. Publications Office of the European Union, Luxembourg (forthcoming).

[23] https://eurocities.eu/

[24] https://living-in.eu/

[25] https://www.ccre.org/

[26] https://citiesfordigitalrights.org/

collectively face key societal challenges, starting from the Smart City Market Place[27] from which this project belongs to.

6. **Act as a trusted data intermediary:** Most of the time, citizens are not able or interested in participating directly in data governance; thus, cities can act as trusted data intermediaries, who manage different strands of citizens' data (including from the private sector or citizen-generated) on their behalf and for the public interest. Local administrations can also put in place initiatives for safeguarding citizens' rights to control and use their own personal data (as in the DECODE project). For successfully achieving these objectives, citizens need to be reassured that the city administration is a trustworthy intermediary. Trust can be increased adopting different strategies, such as public engagement and consultations, or launching civic initiatives aimed at promoting digital rights and co-creating programmes that foster data literacy and data use for public interest (such as based on crowdsourcing data) (Ponti and Craglia, 2020).[28]

[27] https://smart-cities-marketplace.ec.europa.eu/

[28] Ponti M. and Craglia M. Citizen-generated data for public policy, European Commission, 2020 JRC120231. https://ec.europa.eu/jrc/communities/en/community/digitranscope/document/citizen-generated-data-public-policy

14

Data Valuation and Its Applications for Smart Cities

Mihnea Tufiş

Eurecat Technology Center of Catalunya, Carrer de Bilbao, Barcelona
Email: mihnea.tufis@eurecat.org

Abstract

The global economy is increasingly more reliant on data, with businesses adopting data-enabled decision-making practices in the form of analytics or machine learning. Views of data as an asset and the steady emergence of data markets depend on the capability of quantifying the value of data. We argue that the data-as-an-asset approach and focusing on assigning a price tag for data is complicated, due to the properties of data, the multiple value chains that it can generate, as well as legal and ethical implications. We introduce a data valuation process that recognises and integrates the contextual nature of data value, together with data quality and data utility assessments. The value of data is reported in a multi-faceted scorecard, which allows for an exploration of data value at different levels of aggregation. We explore how cities can benefit from the multitude of data they harvest in the process of digitalisation, and we argue that these benefits can be enhanced if cities were to have a more concrete understanding of the value of their data. We discuss their multiple roles with respect to big data processing, as producers, consumers, regulators, and educators for their citizens, and conclude with a list of data-centred actions that cities should implement as part of their smart city agenda.

14.1 Introduction

The Covid-19 pandemic and the current geopolitical changes are bringing a sense of urgency about the actions needed to tackle challenges raised by

DOI: 10.1201/9781003399384-15

older and deeper transformations – climate change, worldwide demographic dynamics, technological advancements, and their impact on the job market.

With increased pressure on central authorities, local governments may just have the opportunity to assume a more hands-on role in dealing with these challenges. Local administrations are best positioned to understand local challenges; they are able to design in terms of local specificities, they can develop a direct link to their citizens and businesses, and they can seek to create local and regional partnerships with other communities that share the same characteristics, challenges, or commitments.

The role of ICT solutions in achieving goals specific to smart city transformation is now indisputable. For example, ICT-based solutions could reduce commuting by 15%–20% and cut greenhouse gas emissions by 10%–15%,[1] and this at a rate of only 12% of city data currently being analysed and used for decision-making and management.[2] A special class is the big data solution, built around a backbone that typically involves massive data collection, processing, and analysis at scale; further downstream, these data are used by more advanced machine learning and artificial intelligence solutions to train autonomous systems for knowledge representation, reasoning, and decision-making. These solutions have the potential to impact every area in the life of cities and their citizens: infrastructure planning, mobility, land and district planning, energy, demographics, social inclusion, development of and engagement with local businesses, culture and entertainment, tourism, health services, and government.

Data scientists know too well that data is key to the success of every solution. Performing a significant analysis or training a reliable decision-making model has instant demands concerning data quantity and quality. And as soon as they step out of open and curated datasets, they quickly run into issues related to availability, interoperability, bias, and many more.

The discussion about data value cannot start without observing the change in the data production–consumption cycle, a consequence of the big data "revolution". Historically, data was produced when it was needed, tailored to the needs of those who would use it, and often consumed precisely by the same actors who enabled its generation. For example, a scientist who

---

[1] Woetzel, J., Remes, J., Boland, B., Lv, K., Sinha, S., Strube, G., Means, J., Law, J., Cadena, A., & von der Tann, V. (2018). Smart Cities: Digital Solutions for a More Liveable Future. McKinsey Global Institute.

[2] Gualtieri, M., & Yuhanna, N. (2014). *The Forrester Wave: Big Data Hadoop Solutions, Q1 2014*. Forrester. https://www.forrester.com/report/The-Forrester-Wave-Big-Data-Hadoop-Solutions-Q1-2014/RES112461

would need to measure the levels of a pollutant in a river would design the data requirements and model, organise, and perform the data collection and eventually process and analyse the collected data. Today, big data refers to massive, continuous, and often loosely structured data, a large portion of which is a by-product of activities, behaviours, or processes that are not always the primarily intended focus of data observations. We have shifted from recording data about a selected subset of activities, to generating data about nearly every aspect of our lives.

The success of a data-centred initiative lies in unlocking the value that data can generate in each context and with respect to the problem being solved. The current paradigm for data production and consumption leads to the perception of data as an asset, subject to exchange, the subsequent appearance of new stakeholders whose activity is based on the acquisition, re-packaging, and selling of datasets, and, finally, the steady emergence of data markets. In this context, a question is becoming increasingly pervasive: *what is the value of my data*?

Thus, the necessity to develop a process for establishing the value of data arises. Ideally, such a process should generalise to any kind of data, application domain, or economic sector. In the context of smart cities, its benefits would be manifold:

- Municipalities could understand the value chains generated by the data they are collecting.
- Consequently, they would be able to map the data value chains to practical outcomes and quantify their impact in the communities.
- Some of these data – depending on ethical implications – could even be exchanged in the emerging data markets and, thus, become a source of revenue for the communities and their individuals.
- A transparent methodology for data valuation could help develop a fair and responsible ecosystem around data markets.
- Participating in the creation of data spaces and digital federations would increase the cooperation between cities within the same region or dealing with similar challenges.
- Educate citizens on the real power of their personal data, its role in today's digital world, and empower them to have more control over this personal aspect of their lives.

In this chapter, we begin a discussion about the value of data: how to define it, what are its main drivers, and why is it difficult to establish it? We also introduce the data valuation process and the data valuation component, the results of a three-year-long R&D project. Then, each section will try to connect to the area of smart cities, by exploring the bidirectional relationship between cities and data value – how each aspect of data valuation can help cities in their digital transformations as well as how cities, through the diversity of the data they collect and challenges that they present, can inform and improve data valuation methodologies.

14.2 Defining the Value of Data

14.2.1 Data through an economic lens – trading data

There is an interesting comparison that is usually made when illustrating the economic value of data and the difficulty of estimating it, especially when perceiving data as an economic asset: "*Facebook is now worth about $200 billion. United Airlines, a company that actually owns things like aeroplanes and has licenses to lucrative things like airport facilities and transoceanic routes between the U.S. and Asia, among other places, is worth $34 billion*".[3]

The view of data as a commodity gathered momentum with the advent of targeted online advertisement and its reliance on personal data. The "classic" model for (personal) data exchanges is for data-centric companies to offer a so-called "free" service in exchange for the users' personal data – the famous "if it's free, then you're the product". With the data deluge from the past decade and the gradual shift of businesses towards data-driven decision making, a gap appeared between their new aspirations and their data know-how. This has created the opportunity for a new group of stakeholders – data brokers – to join in an already unbalanced ecosystem. These intermediary enterprises exist "solely to collect personal data and sell it as a commodity to retailers, advertisers, marketers, even other data brokerages and government agencies".[4]

It is important to make the distinction between personal and non-personal data. According to Article 4 of the EU General Data Protection Regulation (GDPR),[5] personal data refers to information relating to natural

---

[3] Baldwin, H. (2015). Drilling Into The Value Of Data. Forbes. Retrieved from https://tinyurl.com/3jytwus9

[4] Madsbjerg, S. (2017). It's Time to Tax Companies for Using Our Personal Data. The New York Times. Retrieved from https://tinyurl.com/3dcpzrvt

[5] Regulation (EU) 2016/679 of the European Parliament and of the Council – Of 27 April 2016 – On the protection of natural persons with regard to the processing of personal data and on the free movement of such data, and repealing Directive 95/46/EC (General Data Protection

persons who can be directly or indirectly identified from the data in question.[6] Personal data raises many additional challenges of social, legal, and ethical nature: what ownership model to adopt, should we even adopt one – since this would involve selling a form of identity,[7] and how do we adapt to different legal frameworks and different interpretations of privacy across cultures, to enumerate just a few.

Data brokers add value to personal data which individuals generate or release during various online activities, by analysing it, aggregating it, generating user profiles, and enriching it with valuable (and often free) data compiled by the National Statistics Organisations (NSO). It is these bundles of repackaged data that are then sold back to different companies to power their data-centred use cases. The global data broker market size was around $246 billion in 2020 and is expected to grow to $365 billion by 2027.

The most recent evolution of data brokers comes in the shape of online platforms for monetising personal data. These platforms claim to be giving back to individuals the control over personal data and enable them to sell it themselves, ideally, by choosing what data and to whom. There does not seem to be much separating these platforms from large data brokers (and, in fact, there is nothing to prevent such a platform from growing into one), but where they do set themselves apart is that they acknowledge the value of personal data and are open to sharing a piece of the revenues with those who generate it.[8]

Finally, this complex landscape is completed by the presence of governments, trying to find their role within it, depending on their degree of understanding of today's digital transformations. First, governments are expected to assume a regulatory role with respect to data exchanges in general, and data brokers in particular. In Austria, a reported discussion about applying VAT on revenues resulting from big data transactions by social media companies was abandoned, citing difficulties in assigning a value to such a transaction.[32]

Regulation), no. Regulation (EU) 2016/679, European Parliament, 88 (2016). Available at https://tinyurl.com/mr3sxrm5

[6] Such information can refer to "*an identifier such as a name, an identification number, location data, an online identifier or to one or more factors specific to the physical, physiological, genetic, mental, economic, cultural or social identity of that natural person*".

[7] Renieris, E. M., and Greenwood, D. (2018). Do we really want to "sell" ourselves? The risks of a property law paradigm for personal data ownership. Medium. Retrieved from https://tinyurl.com/272uxk78

[8] Tufiş, M. and Boratto, L.. (2021). Toward a Complete Data Valuation Process. Challenges of Personal Data. J. Data and Information Quality 13, 4, Article 20 (December 2021), https://doi.org/10.1145/3447269

Similarly, the United States Senate started holding hearings with respect to the DASHBOARD Act (Designing Accounting Safeguards to Help Broaden Oversight and Regulation of Data), a piece of legislation designed to protect individuals' privacy by forcing companies to disclose to the users the "true value" of the data that concerns them.[9]

Beyond its role as a regulator, there are instances in which governments seek to act as a data broker itself. In 2014, citing the abundance of data it amasses, the UK ministers attempted to pass legislation that would allow HM Revenue & Customs (HMRC) to sell anonymised taxpayers' data to third parties. This came under harsh scrutiny since the British government's track record in terms of data security and data anonymisation practices is far from clean.[10] More worrisome is that despite restrictions and criticism, the HMRC went ahead and quietly released VAT registration data "for research purposes" to three private credit rating agencies (Experian, Equifax, and Dun & Bradstreet).

Establishing an equitable relationship with data brokers will be a challenge for cities in their quest to become smart through digitalisation. Cities may find themselves in a position similar to that of big data companies, in the sense that they are both able to generate as well as consume a large amount of data, some of which is behavioural and often personal. However, as opposed to big tech companies, cities are not primarily run for profit; a city's goal should be the wellbeing of its citizens, and following such principles, it can set it up as a new type of actor in the data exchange landscape – one that generates value through a responsible use of its citizens' data and redistributes this value back to them to improve the livelihood of the community on which it is built upon.

14.2.2 The price of personal data – a chaotic landscape

There is a wide range of personal data collected by data brokers: identification, demographic, location, behavioural, online activity, psychological, product, and political preferences. Most of the times, these data are sold in bundles, which prompts several questions: are all these equally important to a buyer, are they equally sensitive for a seller, and how do each of these stakeholders value them? A reward as low as one cent a month for sharing

[9] S.1951 - 116th Congress (2019-2020): Designing Accounting Safeguards To Help Broaden Oversight and Regulations on Data (2019/2020).

[10] Mason, R. (2014). HMRC to sell taxpayers' financial data. The Guardian. Retrieved from https://tinyurl.com/4x4xcptp

exclusively location data might not convince a user to give it away; however, a bundle of several data types that can amount to as high as $100/month could prompt individuals to invest time in building, managing, and selling personal data portfolios. A second observation concerns the wide price range at which the same type of data is sold. For example, Luth Research pays $100/month for a bundle containing location, social media activity, and browsing activity,[11] whereas Datacoup used to pay $8/month for a similar package.[12] We believe this discrepancy is due to the lack of established data markets, data trading rules, and, as we will see next, a significant gap between the monetary value expected by individuals and what is actually paid by data brokers.

There are also examples of good practices in terms of dealing with user data. Wibson Data Market[13] is trying to enforce transparency, by stating who the data is generated for and for what purpose. Spanish company Telefónica proposes the establishment of a data bank which allows their service users to log all their activities on the network; this is somewhat similar to AT&T's Gigabit, with the major difference being that the former would give users full control over their data,[14] whereas the latter would charge them an extra $29 for keeping their data private.

A study by another telecom giant, Orange, covering 2023 mobile phone users balanced across age categories and countries of origin (France, Poland, Spain, and UK),[15] suggests the existence of three factors that influence the perceived value of personal data:

1. the usefulness of the data to the beneficiary organisation;
2. the type of data;
3. the risk associated with sharing it.

The study also underlines that users are aware that their data is valuable to organisations, which can benefit from it and reveals an ordering relationship of how likely they are to share types of personal data (demographic ≻ activity

[11] Ross, W. (2014). Is Your Smartphone Privacy Worth $100 a Month? MIT Technology Review. Retrieved October 18, 2019, from https://tinyurl.com/mrv9czyp

[12] Simonite, T. (2013). Coming Soon: Take Your Own Personal Data to Market. MIT Technology Review. Retrieved from https://tinyurl.com/ksf68nwa

[13] Travizano, M., Sarraute, C., Ajzenman, G., & Minnoni, M. (2018). *Wibson: A Decentralized Data Marketplace* (arXiv:1812.09966).

[14] At the time of submitting this text, there are no mentions as to what the price of such a service would be and how would Telefónica benefit from it.

[15] Loudhouse. (2014). The Future of Digital Trust. A European study on the nature of consumer trust and personal data (Industry No. 2; The Future of Digital Trust, p. 7). Orange.

and behavioural ≻ third-party or financial data).[16] The study also points to a paradox in consumers' understanding of sharing personal data: while a majority of respondents (77%) declare that privacy and transparency of data usage are important and identify the risk attached to sharing as an important factor influencing data value, they also indicate demographic data as the type they would most likely share – despite the clear risk of identity theft and online fraud attached to it.[17]

In a 2016 survey, credit comparison site Totally Money[18] asked 1000 UK consumers to estimate[19] the economic value of different categories of personal data. The results revealed interesting attitudes and different data-sharing practices, spread across demographic groups and types of data alike: young people (18–24 years old) value their data the most, while millennials value theirs the least (£1773); men value data about their online activity higher than women do (£1112 vs. £859 for email data, £1056 vs. £817 for browsing data, and £951 vs. £778 for location data). Perhaps the most surprising result of this study is the difference between the average self-estimate of respondents' personal data (£2031) and how much brokers are paying for it (£0.45). While the methodology of this study is not completely clear and it is difficult to assess the representativeness of the sample, the magnitude of this difference, together with the paradox observed by Smith,[17] points to the necessity of building "digital literacy", together with legal frameworks suitable to the consequences of the permeability of our digital traces and to the ease with which data companies can process and monetise them.

Interesting results are also coming from academia, with a recent increase in the study of methods for valuing user-generated data, particularly geolocation and online behaviour. In one of the most relevant experiments, Staiano *et al.*[20] simulated a data market for personal data transactions. Participants

---

[16] Third party data: email, personal preferences of other contacts; Behavioural data: location, mobile purchase history; Demographic data: name, date of birth, phone number.

[17] Smith, M. (2016). Proximus starts selling customer data reports for €700 a time. European Communications. Retrieved from https://tinyurl.com/muwm2ncd

[18] Davies, J. (2016). Consumers price their data at £2k – Companies pay 45p. Telecoms. Com. Retrieved from https://tinyurl.com/ysta7zwb

[19] TotallyMoney.com conducted research in June 2016 to identify the prices third-party companies pay for data to utilise in marketing campaigns: Financial Times, The Telegraph, McAffee, CostOwl.com, OnePoll.com.

[20] Staiano, J., Oliver, N., Lepri, B., de Oliveira, R., Caraviello, M., and Sebe, N. (2014). Money Walks: A Human-Centric Study on the Economics of Personal Mobile Data. Proceedings of the 2014 ACM International Joint Conference on Pervasive and Ubiquitous Computing - UbiComp '14 Adjunct, 583–594.

were equipped with devices gathering various types of data (calls, applications usage, location, and media usage) at three levels of aggregation (individuals, processed, and aggregated). They were then expected to sell the data to the Research Laboratory during auctions (reverse second price[21]), initially running weekly and then daily. In the nearly 600 auctions organised, participants received rewards totalling approximately €270, with a median price of €2 across categories. The auctions were also able to cast a light into the self-valuation of personal data, and just like in the study conducted by Orange,[22] it revealed an order of perceived value among the data types: location ≻ communication ≻ apps ≻ media; not surprisingly, processed data was held to a higher value than raw data. Two additional observations may be important take-aways when designing data valuation methods:

1. Increasing the frequency of auctions (from weekly to daily), decreased the value of the bids; an indication that the data market may play by the rules of supply and demand.
2. The value of data increased when unexpected situations arose (traffic jams caused by either a weather event or a local holiday); this suggests that the value of the same data is highly dependent on the context.

14.2.3 Challenges defining the value of data – beyond financial value

Until now, attempts at establishing the value of data were connected to large impact business events – mergers and acquisitions, bankruptcy, data transactions, and data breaches – and usually focused on the monetary value of data. This is perhaps why comparisons between data and other commodities (oil, gold, etc.) are usually making media headlines or are thrown in as a hook in conversations on the topic. We generally understand that there is value in data, mostly by connecting its applications to immediate outcomes and benefits, but it is very unclear what the source of this value is, the mechanisms through which it is created, and what it consists of (beyond the obvious monetary aspect).

Short and Todd[23] consider the value of data as the composite between the value of the asset itself, the value resulting from its use, and its expected or

---

[21] The lowest bidder wins, but the reward will equal the second-lowest bid.

[22] Loudhouse. (2014). The Future of Digital Trust. A European study on the nature of consumer trust and personal data (Industry No. 2; The Future of Digital Trust, p. 7). Orange.

[23] Short, J. E., and Todd, S. (n.d.). What's Your Data Worth? MIT Sloan Management Review. Retrieved November 8, 2020, from https://sloanreview.mit.edu/article/whats-your-data-worth/

future value. In her comprehensive inventory of academic papers and industry reports on the value of data, Slotin[24] observes how "striking [it is] that among [the] diverse perspectives, each author is grappling in their own way with the implications of data as a new economic asset, and yet there appears to be little consensus on how best to measure its value. One thing they can agree on is that measuring the value of data – and making [a] case for investing in data – is very difficult". Analogies with either tangible (oil) or intangible assets (patents, intellectual property, etc.) break at the point where the mapping between properties and assigned value becomes less clear (e.g., what is the difference in value between 32 and 35 GB of the same data? What is the difference between data that is 55% and 65% accurate? What is the value of a dataset that has already been used to train a machine learning model?). And perhaps this is normal since rules that apply to old commodities possibly do not even apply to this new kind of resource. To understand the difficulties of assessing the value of data, Mawer[25] follows the progression through each element of the data value chain, from raw data to action and potential value, and maps them to the sequential stages of the data lifecycle (discover, ingest, process, persist, integrate, analyse, and expose).

This follows on work by Porter[26] and Kaplinsky[27] – the first to describe value chains applied to the design, production, and delivery of products and services – subsequently adapted into knowledge value chains.[25,28] To explain the value of knowledge co-production, Peppard and Rylander[29] needed to break the linear model. They introduced the concept of network value, which allows its participants to function independently, within a framework of common principles.

Attard *et al.*[30] recognise the non-linearity of data value creation and refine the previously presented work into the data value networks (see Figure 14.1).

[24] Slotin, J. (2018). What Do We Know About the Value of Data? Global Partnership for Sustainable Development Data.

[25] Mawer, C. (2015). Valuing Data is Hard. Silicon Valley Data Science. Retrieved from https://tinyurl.com/495mt343

[26] Porter, M.E.. (1985). Competitive Advantage: Creating and sustaining superior performance. NY: Free Press.

[27] R. Kaplinsky, R. and Morris, M. (2002) A Handbook for Value Chain Research.

[28] Lee, C.C. and Yang, J. (2000). Knowledge value chain. Journal of Management Development, 19(9):783–794, 2000.

[29] J. Peppard and A. Rylander. (2006). From Value Chain to Value Network: European Management Journal, 24(2-3).

[30] Attard, J., Orlandi, F., & Auer, S. (2017). Exploiting the Value of Data through Data Value Networks. *Proceedings of the 10th International Conference on Theory and Practice of Electronic Governance*, 475–484. https://doi.org/10.1145/3047273.3047299

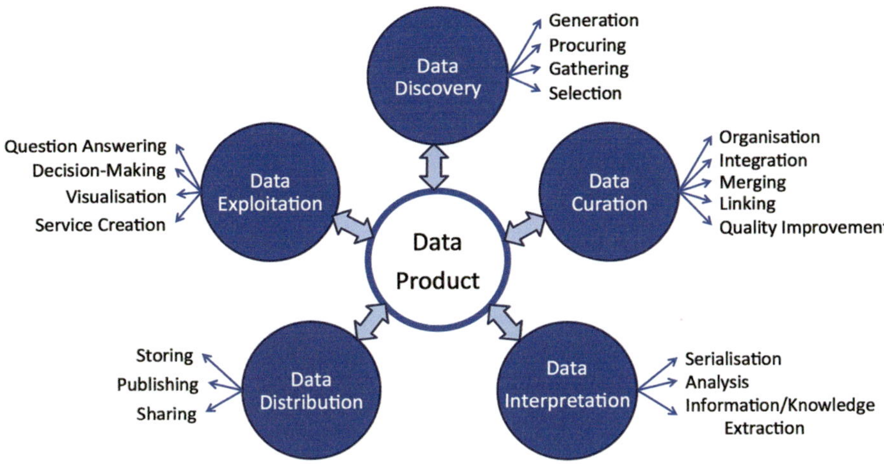

Figure 14.1 Data value network as illustrated by Attard *et al.*[30]

This is composed from activities (depicted by circles), each of them consisting of several actions (connected to the circles by simple arrows).

The strength of the network lies in the fact that it is built precisely around the characteristics that make the value of data difficult to assess:

1. Non-tangible product. It is infinitely shareable, and while some data may deprecate with time, it can be reused over and over, without losing its properties. It is also true that in some cases (e.g., social media big data, a company's financial data, industrial production data) part of these data's value relies on their uniqueness or on their sharing restrictions. However, excepting these cases, given the promotion of FAIR principles[31] and open science, accessibility (as opposed to closed silos) is expected to increase its value.

2. Non-sequential processing. While theoretically data processing pipelines have a relatively linear progression, in practice, things are a bit more complicated. Value generating activities can be skipped, executed in parallel, or in slightly different order than the theoretical one, or be a part of iterative loops.

3. Several actors can cooperate when realising the activities. Moreover, each of these actors may output a data product on their own or contribute to the co-creation of a data product.

[31] https://www.go-fair.org/fair-principles/

4. The network allows for the existence of nested value chains. These are formed by the actions that compose each activity.
5. The network allows for the existence of recurring value chains. This implies that value chains can be created as long as the data stays relevant. The data product resulting from a certain activity could be the "final product" or could form the input to a branching activity, thus perpetuating the chain.
6. The activities can be performed independently by any number of actors.

This model currently appears to be the best equipped to model and possibly quantify the high context-dependent nature of the data valuation process.

14.3 The Data Valuation Process

The data valuation process (DVP) and the data valuation component (DVC), implementing it, were developed as part of the Horizon 2020 Safe-DEED (Safe Data Enabled Economy Development) project.[32] It considers that the value of data is generated from two main areas: data quality and data usability, which are assessed through the lens of the context in which the data will be used. The context is set by the user, during a context definition procedure, based on which the relevant components of data quality and data usability are established (see Figure 14.2).

The tool is trying to maximise the automation degree of all these processes and proposes in-depth analyses to support the value of data and the reduction of the time dedicated to the data valuation process.

Since this is a complex problem, the presentation of the results avoids the generation of a single aggregate value. Instead, the platform generates a set of scores (for different perspectives and at different levels of detail), thus informing the user on the strengths and weaknesses of the dataset they are assessing.

Next, we look at each of the building blocks of the DVP and we try to understand how cities can use them (either as part of the DVP or as independent components) to understand and properly use the value of their data.

14.3.1 Data contexts

Let us consider a dataset containing GPS traces of taxis in a city. For a ride-hailing application, such data would provide a way into estimating the

[32] https://safe-deed.eu/

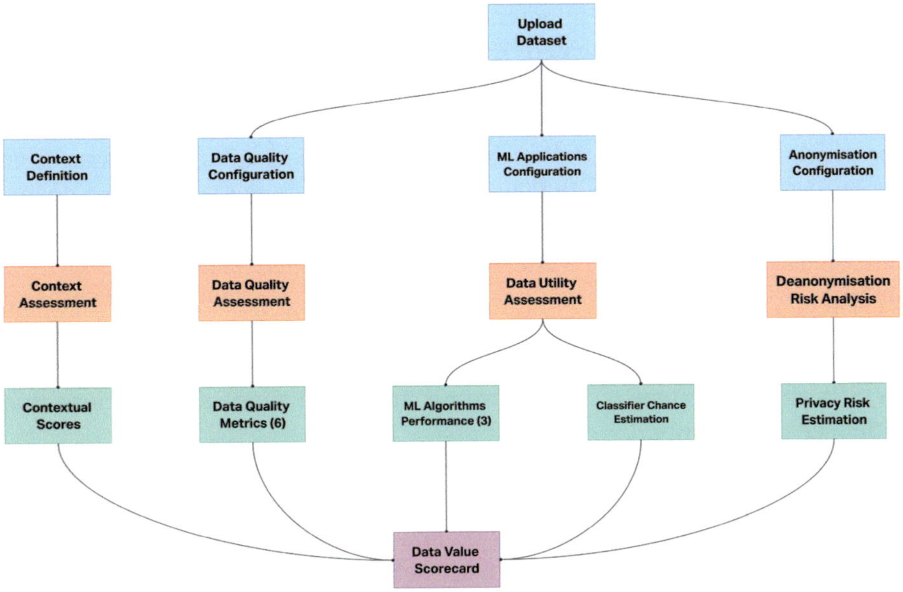

Figure 14.2 Overview of the components of the data valuation process. (Components requiring interaction with the user are in blue (rows 1 and 2 from above), automatised components in orange (row 3), scoring components in green (row 4), and the output component in purple (row 5). The arrows indicate the sequence of actions from processing the data to generating the result.)

customer needs in different areas of the city, at different times, allowing them to develop machine learning solutions for load balancing and trip planning and eventually maximising their revenues. The local administration could use this data to understand road congestion and travel times and plan infrastructure interventions (repairs, extensions, and restrictions), modify public policies (congestion taxes), or plan connected services (public transportation). A retailer could look at this data in conjunction with other sources and understand behavioural patterns of people living in different areas of the city and thus plan opening schedules, logistic operations, or decide to open new branches.

Data can have different values for different roles within the same organisation. Data containing the flow of passengers through a certain area might be enough for a planning manager and his team who decide to build a larger shelter or a new bus stop, but for the R&D department working on a new routing algorithm, such information might be overly aggregated and useless for their necessities. Even within the same department, different tasks might impose different requirements from the same dataset. The data science team might be able to provide a good enough analysis of travel patterns from data which contains trips aggregated over 30-minute intervals, but such a dataset will not be useful if the task is to create an accurate traffic prediction model.

The contextual nature of data is often cited as one of the main reasons for which assigning value to it is difficult. We have seen how different value chains can be completed with the same raw data or how similar value chains can be completed with different raw data,[24,29] depending on the purpose of the data processing. Slotin[23] extends that observation and concludes that context-specific, impact-based methods might be the most suitable for communicating data value, despite this specificity also being their main drawback. In their data quality principles, the US National Institute of Statistical Science (NISS) cite contextual factors (purpose, user, and time) among those that influence data quality.

Building a solution that takes contexts into account has, first, to surmount the challenging aspects of defining, formalising, and encoding them. With research focused specifically on contexts for data value being almost inexistent, we turned to the literature on metadata for datasets and data quality assessments to seek for meaningful parallels. This confirms the context-dependent nature of data value and brings a first level of clarity concerning the layers that compose a valuation context:

- organisational profile;
- business user profile;[33,34,35]
- a specific task, personal preferences;[34]
- business rules/processes;[32]
- organisational and government regulations.[32,33]

In a sense, defining contexts is akin to understanding users, identifying use case scenarios, and deriving user requirements.

Recent work focusing on data profiling and valuation of metadata offers valuable leads into how data valuation contexts could be established and

---

[33] Pipino, L. L., Lee, Y. W., and Wang, R. Y. (2002). Data quality assessment. Communications of the ACM, 45(4), 211–218.

[34] Cai, L., and Zhu, Y. (2015). The Challenges of Data Quality and Data Quality Assessment in the Big Data Era. Data Science Journal, 14(2), 1–10.

[35] Even, A., and Shankaranarayanan, G. (2006, November 10). Value-Driven Data Quality Assessment. Proceedings of the 2005 International Conference on Information Quality. MIT IQ Conference, MIT, Cambridge, MA, USA.

quantified. Among them, we distinguish a questionnaire-based method for mapping data properties to data value,[36] the creation of datasheets for datasets,[37] and the Dataset Nutrition Label,[38] a diagnosis framework providing critical information at the point of data analysis.

As for the DVP, it requires that a user provides as much information as possible about the context in which a dataset will be used. This is done through a questionnaire with clear answers, which are then mapped to values, yielding the contextual value of the dataset. The questionnaire is structured in the following layers:

1. Systems and economics: availability and access; purpose
2. Legal and obligations: data protection; legal-terms-obligations
3. Data science: tools; format
4. Data properties: data velocity; data transformations; data quality; data age
5. Business: frequency of use; benefits

These early methodologies for context formalisation underline the importance of metadata that accompany a dataset. Metadata give a generic view into the origins of a dataset, the methods for generating it, the purpose for which it was generated, its format and access to it, the licenses that may apply to it, and the methods and tools used to process it up to its current form. Many of the data-driven stakeholders, whether producing or consuming data, are far from insuring the bare minimum with respect to metadata. This results in data being difficult to index, find, and (re-)use, essentially decreasing or even cancelling its value.

Through their technology departments, we recommend that cities step forward and assume the responsibility for creating such metadata. For those cities that already have IoT and e-government infrastructures in place, documenting such metadata (either by filling the DVC context valuation forms or

[36] Kannan, K., Ananthanarayanan, R., and Mehta, S. (2018). What is my data worth? From data properties to data value. http://arxiv.org/abs/1811.04665

[37] Gebru, T., Morgenstern, J., Vecchione, B., Wortman Vaughan, J., Wallach, H., Daumé III, H., and Crawford, K. (2020). Datasheets for Datasets. http://arxiv.org/abs/1803.09010

[38] Holland, S., Hosny, A., Newman, S., Joseph, J., and Chmielinski, K. (2020). The Dataset Nutrition Label: A Framework to Drive Higher Data Quality Standards. In D. Hallinan, R. Leenes, S. Gutwirth, and P. De Hert (Eds.), Data Protection and Privacy: Data Protection and Democracy (pp. 1–26). Oxford: Hart Publishing.

creating datasheets[36]) would add value to their current efforts and allow them to explore further avenues for their data.

For those cities that are still at the beginning of their digital transformation journey, this may be a good opportunity to do things right from the onset, by creating the data collection and management infrastructure, processing pipelines and metadata in a coherent way.

At this point, it is crucial that cities reach out to data and information practitioners, since these communities have the know-how needed to build fair and efficient data and metadata infrastructures. The ideal team should include:

1. Technical experts: data scientists, data engineers, library, and information science professionals.
2. Legal experts, preferably specialised in technology, intellectual property, or consumers law, able to advocate for the citizens who will be impacted by data processing, and foresee or react to future ethical and legal issues that will arise.
3. Experts from within city administration, preferably project managers who are able to map the requirements derived from urban challenges to technical solutions. These specialists should be the glue connecting the needs of the city and its citizens, the technical solution, within the legal and ethical boundaries.

While building an in-house team would be the ideal setup, budget limitations might require cities to seek for partnerships with universities, technology centres, or private companies. In these cases, cities should proceed with care, as lack of budget, technical knowledge, or legal safeguards may lead to their data being used for other purposes than those intended. Losing the trust of the citizens (either as data subjects or users of data products) can lead to the failure of digitalisation efforts altogether.

The creation of metadata and their evolution towards a standard for context formalisation is crucial for the success of data valuation methodologies. They provide the building pieces for describing a variety of contexts and discovering those dimensions that are important for the value of data in these contexts. Generalising the different contexts would enable for both context-dependent and context-independent analyses of the value of data. Finally, connecting these contexts to quantifiable values can lead to the establishment of transparent and fair data markets.

In this respect, cities have the advantage of processing a wide variety of data (both personal and non-personal), which makes them the ideal partners for pushing the R&D of methods for data valuation.

14.3.2 Data quality assessment

The earliest preoccupations towards a formal understanding of quality date back to its application to assembly-line production and manufacturing in the beginning of the 20th century and accelerated later in the 1950s and 1970s with its adoption to business practices. Along the years, various definitions have been put forth, referring to quality as "conformance to requirements",[39] Joseph Juran's famous "fitness for use",[40] or the "degree to which a set of inherent characteristics fulfils requirements".[41] One definition refers to quality as the "value to some person",[33] recognising the intrinsic value derivable from data quality as well as its contextual nature.

With the development of ICTs, interest in quality of data has sparked during the 1990s. The democratisation of the internet and the advent of big data and data-centred solutions generated more interest in the topic and laid the ground for a currently mature and dynamic research field. In 1996, the Total Data Quality Management Group at MIT adopted the "fitness for use" definition and acknowledged its dependency on the consumers. The principles of data quality by the US National Institute of Statistical Sciences (NISS) adopt the view of data as a product and, as such, consider that its quality results from the process that generates them. Later, data quality was enacted at the governmental level, as was the case of the US Data Quality Act[42] or the Welsh Data Quality Initiative Framework.[43] In Europe, Bergdahl *et al.* report on the successful integration of data quality assessment in the activities of several National Statistics Organisations: Statistics Sweden, Statistics Norway, CBS in the Netherlands, the Austrian Quality Concept (an in-house quality reporting system), the ONS Guidelines for Measuring Statistics Quality (a grading scheme for statistical products), and the Slovenian Statistical Office (data quality measurement for short-term statistics).[44]

[39] Crosby, P. B. (1988). Quality is Free: The Art of Making Quality Certain, New York: McGraw-Hill.

[40] Juran, J.M. (1951). Quality Control Handbook. 4th ed.

[41] International Organisation for Standardisation. (2015). ISO 9000 Family for Quality Management Systems.

[42] Office of Management and Budget. (2006). Information quality guidelines for ensuring and maximizing the quality, objectivity, utility, and integrity of information disseminated by agencies. Available at: https://tinyurl.com/ychxhmsd

[43] NHS Wales. (2004). Data Quality Initiative Framework. Project Report.

[44] Bergdahl, M., Elvers, E., Földesi, E., Kron, A., Lohauß, P., Mag, K., Morais, V., Nimmergut, A., Viggo Sæbø, H., Timm, U., and Zilhão, M. J. (2007). Handbook on Data Quality Assessment Methods and Tools. European Commission - Eurostat.

Data quality can be regarded as the ability of data to serve its purpose – generally seen as the needs of an organisation in terms of operations, planning, and decision-making.[45] Therefore, in order to evaluate the quality of data, a plethora of data quality assessment methodologies have been developed over the recent years, adopting different perspectives and covering an even larger spectrum of quality dimensions in their attempt to encompass the multitude of assessments that gather under the data quality umbrella.

To clarify, "a Data Quality Dimension (DQD) is a recognised term used by data management professionals to describe a [property] of data that can be measured or assessed against defined standards in order to determine the quality of data".[46] Dimensions focus on measuring and communicating the quality of data, as opposed to describing what the data represents.

14.3.3 Data quality metrics and dimensions

Historically, there is a correlation between the development of ICTs and that of data quality assessment methods. The early systems were monolithic, usually consisting of a single data source and simple data flows, and the only source of errors would come from data entry. Data quality would, therefore, involve accuracy, consistency, completeness, and time-related metrics. The evolution towards network-based systems involved a re-adaptation of these dimensions; with the later advent of the web, data sources have become more numerous and more varied and, as a consequence, new dimensions such as accessibility and reputation had to be considered. Currently, peer-to-peer systems require a new rethinking of these dimensions and, more importantly, the consideration of privacy issues. This evolution of ICT systems is itself one of the causes for the number of methodologies, some of which specialised on subsets of data quality issues.

An overview of all dimensions and subsumed metrics[47] allows us to confirm the complexity and multi-dimensionality of the concept of data quality. The Total Data Quality Management Group at the MIT defines 15 quality

[45] Lebied, M. (2018). Guide To Data Quality Management & Metrics for Effective Data Control. Datapine. Retrieved from https://tinyurl.com/2p8fmxrx

[46] Askham, N., Cook, D., Doyle, M., Fereday, H., Gibson, M., Landbeck, U., Lee, R., Maynard, C., Palmer, G., and Schwarzenbach, J. (2013). The Six Primary Dimensions for Data Quality Assessment—Defining data quality dimensions. DAMA UK.

[47] Batini, C., Cappiello, C., Francalanci, C., and Maurino, A. (2009). Methodologies for Data Quality Assessment and Improvement. ACM Computing Surveys, 41(3).

dimensions,[48] the Data Management Association for the UK focuses on 6 primary dimensions,[46] and Statistics Netherlands mentions 49 factors that influence the quality of secondary data and groups them into 5 focus areas.[49]

Due to the contextual nature of data quality assessment, there is little to no consensus as to what might be a subset of necessary data quality dimensions to consider. But is there a subset of "basic" dimensions and metrics that should always be considered when assessing data quality?

A review of DQA methodologies points towards a set of four such DQDs, namely: completeness, validity, accuracy, and timeliness.[33,34,44,45,50,51,52,53] ISO/IEC 25012 confirms these as well as "credibility" as inherent characteristics of data quality.[54]

Data quality assessment is key to unlocking the value of data and if they are to embrace digital transformations, cities should place it at the centre of their technical activities. Thus, cities should seek to ensure the assessment along the five DQDs previously mentioned, by using appropriate metrics (see Table IX in the work of Batini *et al.*[47] for a complete list of DQMs). The next natural step would be to invest in the technical expertise necessary to address the shortcomings identified along each of these dimensions. Again, while data quality requirements might differ on a case-by-case basis, there are minimum data quality requirements that are expected from data and seeking to achieve these already increases its general usability.

---

[48] Warner, M.R., and Hawley, J. (2019). Designing Accounting Safeguards To Help Broaden Oversight and Regulations on Data. Retrieved from https://tinyurl.com/3fujstf3

[49] van Nederpelt, P., and Daas, P. (2012). 49 Factors that Influence the Quality of Secondary Data Sources. In: Quality and Risk Management (12). Statistics Netherlands. The Hague.

[50] Behkamal, B., Kahani, M., Bagheri, E., and Jeremic, Z. (2014). A Metrics-Driven Approach for Quality Assessment of Linked Open Data. Journal of Theoretical and Applied Electronic Commerce Research, 9(2), 11–12.

[51] Görz, Q., and Kaiser, M. (2012). An Indicator Function for Insufficient Data Quality – A Contribution to Data Accuracy. In H. Rahman, A. Mesquita, I. Ramos, and B. Pernici (Eds.), Knowledge and Technologies in Innovative Information Systems (Vol. 129, pp. 169–184). Springer Berlin Heidelberg.

[52] Piprani, B., and Ernst, D. (2008). A Model for Data Quality Assessment. In R. Meersman, Z. Tari, and P. Herrero (Eds.), On the Move to Meaningful Internet Systems: OTM 2008 Workshops (Vol. 5333, pp. 750–759). Springer Berlin Heidelberg.

[53] Sebastian-Coleman, L. (2010). Data Quality Assessment Framework. The Fourth MIT Information Quality Industry Symposium.

[54] International Organisation for Standardisation. (2008). ISO/IEC 25012:2008 Software engineering – Software product Quality Requirements and Evaluation (SQuaRE) – Data quality model.

To take on this challenge, cities would need to employ the same type of profiles as in the case of metadata generation. In fact, data quality assessment is an even higher technical endeavour, requiring a variety of profiles: analysts, data scientists, data engineers, domain experts, and legal experts. Therefore, the discussion in Section 14.3.1 about how such a team would work and the comments with respect to outsourcing these activities apply here as well.

Using a solution like the DVC can speed up the process, due to its configurable data quality assessment module. This would not require an entire data science team; it would, however, require the collaboration between a domain expert and a data specialist, who would be able to understand the quality requirements of a dataset and their connection to the problem. Once such experts become trained in using the DVC, they can apply it to any available dataset and, thus, get in-depth knowledge of the quality of data and eventually quantify its value.

14.4 Aggregating and Reporting the Value of Data

The success of the data valuation process depends on its adoption by data practitioners which, given the multi-dimensional nature of data valuation, depends on:

- the capacity of the platform to promote the transparency of the assessment processes;
- the interpretability and replicability of results;
- the degree to which such results can be used by practitioners.

The last item refers to the necessity to aggregate the results of the sub-processes that compose data valuation into a single measure that can be easily understood at different levels of organisations and based on which, ultimately, decisions of economic nature can be made. Thus, notions such as "energy label" for data or "price tag" for data are appealing, especially to those operating at commercial or executive levels of organisations. However, such aggregate measures are both difficult to construct (at least for now) and may lead to confusing or inaccurate interpretations, which could undermine the whole data valuation effort.

Interest in developing a single measure to characterise data first appeared in the community of data quality practitioners. Pipino *et al.* point to the fact that a single-value aggregated measure – a quality index – could be subject to the same deficiencies that affect other commonly used indices (Dow Jones Industrial Average, Consumer Price Index, etc.).[33] These derive from the statistical methods used for estimations, the interpretation of different components, their contribution weight to the final index, the methodologies used for choosing these components, etc. Similar shortcomings are noticed by Bronselaer *et al.*[55] who warn about the difficulty in interpreting an aggregation of DQMs, each referring to very different quality facets. Even when choosing a reporting scale, both groups of researchers point to relevant challenges, whether it is the difficulty of aggregating DQMs operating on different scales[33] or the loss of interpretability of a result that standardises all DQMs in the [0,1] interval.[54] Bergdahl *et al.* mention that previous attempts to compile composite indicators for data quality by NSOs have failed and refer to the contextual nature of DQA as the main constraint for selecting the right subset of indicators and assigning them suitable weights.[43]

Reporting is paramount in promoting the adoption of innovative platforms, especially if they involve complex evaluation processes, like the DVC.

A first component of reporting is data profiling, which is usually performed as an entry point to data quality management,[44] right before data analysis. This gives an initial insight into the data (ranges, distributions of attributes, pair-wise correlations, etc.) and supports the definition of data quality requirements.[56]

Once DQA is performed, there are several approaches to reporting an often-multi-dimensional result and eventual aggregates.

- Report the cost associated with poor quality of data and summarise it in a data quality scorecard.
- Issue a certificate of data quality or a quality alert, depending on whether quality requirements are satisfied or not. It is recommended that only a small number of self-explanatory labels (e.g., “sufficient quality”, “experimental data”, etc.) are created and that, once introduced, they stay in circulation for some time. Labels should also include “expiration dates” and allow for constant recertification of datasets, reactive to changes in content or requirements.
- Create data narratives that highlight the impact of good or bad data. Impact-based approaches for data valuation[52] tell compelling stories and connect data to contexts and clear outcomes.[57]

---

[55] Bronselaer, A., De Mol, R., and De Tre, G. (2018). A Measure-Theoretic Foundation for Data Quality. IEEE Transactions on Fuzzy Systems, 26(2), 627–639.

[56] Jones, D. (2016). Data Profiling vs Data Quality Assessment – Let’s Explain The Difference. Data Quality Pro. Retrieved from https://tinyurl.com/2swe9ww3

[57] Hammond, K. J. (2013). The Value of Big Data Isn’t the Data. Harvard Business Review.

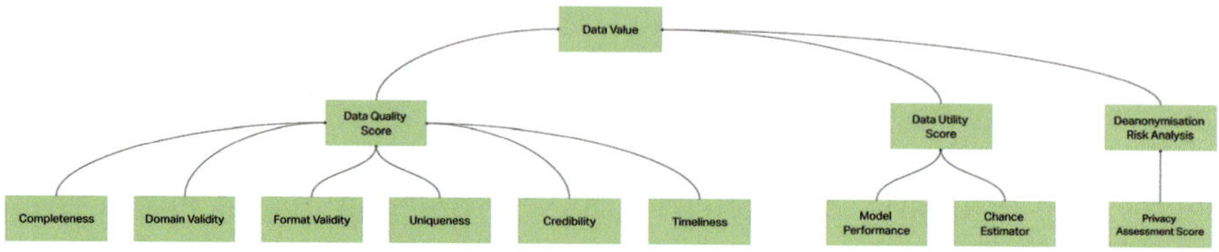

Figure 14.3 The aggregation of the scores generated from the different sub-components of the DVC into the data value scorecard.

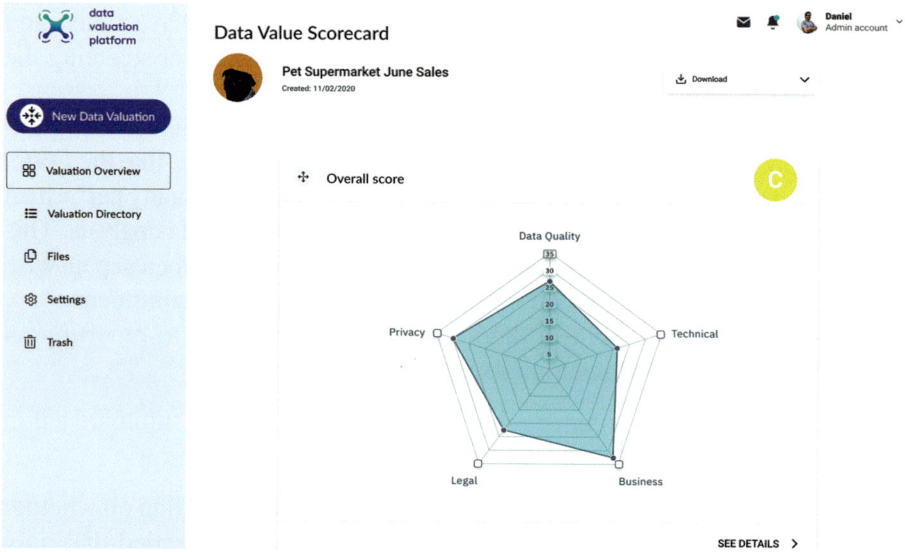

Figure 14.4 Data value scorecard. A combined view over the general data quality score and its composing data quality metrics.

It is important to underline that the DVP does not generate a price tag for data. We have already discussed the technical difficulties for achieving that (e.g., the properties of data as an asset, the properties of data value chains, and lack of adapted economic methodologies). Beyond these, data pricing (and, in particular, private data) raises important legal issues (ownership, intellectual property, etc.), as well as moral ones – should we even engage in transactions and monetary exchanges involving digital extensions of human identities?

The DVP thus focuses on giving a multi-faceted quantification of the components of data value (data quality, data utility, and privacy), within the defined context (see Figure 14.3, Figure 14.4 and Figure 14.5). This

Figure 14.5 Data value scorecard. A multi-dimensional view of the contextual score.

multi-dimensional reporting system allows for stakeholders to grasp the value of data from a variety of perspectives and at different aggregation levels; it does not exclude a one-value score, but it invites practitioners to explore the reasons behind it.

The reporting of data value is a challenging problem, as it tries to connect various dimensions of data value, to valuation contexts, the needs of various stakeholders, as well as human factors. It would greatly benefit from applying user experience design principles to identify the best way to interact with the user and present the outcome of data valuation.

Once again, cities and their communities can prove very useful in helping refine this aspect of data valuation. The variety of challenges, data, and data-centred use cases that cities generate places them as an important contributor to data valuation R&D. Cities can become a facilitator for communities and individuals to have an active role in the processes of digital transformations. Administrations can reach out to their citizens and businesses and organise focus groups, in which they can connect with data valuation professionals. We believe that such exchanges will benefit all parties.

- Data value professionals can educate the large public on the value of data, challenges, and how they can benefit from it. This should promote the use and further development of methodologies and tools for data valuation.
- Cities can discuss with communities their different data-centred projects and try to understand future lines of development and shape public policies. This could be a part of the constant dialogue required between city administrations and citizens.
- Citizens and businesses will be able to communicate their needs and their expectations and help shape the next generation of technologies, which have to become more than harvesting tools of digital identities.

14.5 Takeaways for Cities

Cities are a diverse concentration of people and activities giving rise to a multitude of daily challenges in their everyday functioning, as well as in their quest to serve their communities. Their response to today's societal challenges places them at the centre of digital transformations. The adoption of IoT technologies and the permeability of social media mean that we now have a better view than ever into the lives of these macro-organisms. Obviously, this turns them into ideal testbeds for new, impactful technologies, but it depends on each city (both community and management!) to leverage these characteristics and place cities in the driving seat of these transformations.

Educate citizens about digital transformations and its value for the community. Understanding the potential of the data that is generated in a city must begin with its managers. They are in charge in setting a city's policies and digital education cannot be ignored anymore. From here, this should be diffused towards their communities, by means of communication and education projects aiming to make everyday citizens aware of the capacity of the city to record human behaviour, the variety of this behaviour, the fact that this behaviour generates value and can lead to progress in theirs and other's communities, and the potential misuses of this data and how it can be prevented.

Involve citizens when designing solutions. Citizens need to feel that they are more than just "users" or "data points" and it is up to the cities and their democratic, participative processes to give citizens a voice. Involving citizens in designing technical solutions has multiple advantages.

- First, it will lead to higher adoption of these solutions, especially if these involve advanced technologies. Public funds must go in carefully designed citizen experiences, as opposed to expensive apps with stale designs that nobody uses.
- It will promote trust in these technologies, also improving adoption rates and building an honest relationship between citizens and managers. Cities should avoid becoming yet another big data processor or data broker.
- Finally, it will create a sense of community, which gives more responsibility to the citizens: they care more about their neighbourhood or city, which has benefits in terms of creating safe and resilient communities.

Design fair, explainable, and privacy preserving technology. We are at a point at which artificial intelligence solutions are being deployed at an accelerated rhythm. Nevertheless, the research community is revealing increasingly more cases of bias present both in AI systems and the data that is powering them. Moreover, the collateral effects of some of these systems have been challenging ethical and societal principles, as well as our current legal frameworks. Cities can respond to these by promoting partnerships with researchers from these areas. The trove of data they are holding would do the following:

- Promote partnerships with researchers from these areas. The trove of data that they hold contains a variety of sensitive information (demographic, behavioural, financial, etc.) and could allow specialists to advance the state-of-the-art in data privacy and algorithmic fairness, while cities would benefit from fair, explainable, and privacy enhancing technologies.
- Lead the discussions about the adaptation of legal frameworks. Lawmakers need to work with technical and legal experts on understanding the consequences of deploying data-centred technologies in cities, their potential conflicts with human and citizen rights, and how to transpose this into law. This is particularly interesting in the case of

cities becoming active as data brokers or managing personal data on behalf of citizens.

Get involved in data markets. One possible avenue is for cities to try to monetise the data they generate, by participating in data markets. An interesting discussion concerns the way in which the revenue obtained from such data would be distributed: to every individual as a form of pay-off for the contribution of their data, redirected towards the city's budget and reinvested in local projects or a combination of the two. Obviously, such an avenue would need to address issues such as data ownership or data rights management and would absolutely need to involve the use of privacy and fairness checkups, as described earlier.

Open the data. Don't silo the data! We have discussed how different value chains, executed by different actors, can generate value from data. Opening the data (e.g., by implementing FAIR principles) will tap into the creativity of other stakeholders and support innovations that were not initially considered. Like before, this should be done while previously deploying privacy preserving and fairness promoting mechanisms. Opening personal or any kind of sensitive data has to lead to scientific advancements and the creation of fair, responsible data products, without the cost of exploiting the lives of those who helped generate this data.

Build data teams and know-how. As we mentioned in Sections 14.3.1 and 14.3.3, cities should consider the creation of data-focused technical teams. These teams should be dedicated to the creation of metadata (to facilitate the creation and quantification of data contexts) and the highly technical data quality analysis. Alternatively, cities can opt for using advanced analyses tools, like the DVC, in which case they should invest in preparing technical staff that is able to understand use cases, translate them to technical requirements, and interpret multi-dimensional output. If these activities end up being performed in partnerships with external parties (tech companies, consultancy companies, and universities), they should make sure that they put in place the right legal and contractual mechanisms for protecting the intellectual property of the data and the privacy of the data subjects. Finally, it is highly recommended that cities allocate resources for two important roles:

- technical facilitator, to bridge the gap between city-specific requirements and teams working on technical solutions;
- legal and ethical experts, dedicated to identifying the challenges put forth by data-centred solutions, their impact on citizens, and how they could lead to an evolution of the current legal framework.

Communicate. Data value is a complex, multi-dimensional concept. While everyone is aware of the value of data, this is still difficult to quantify and report. Promoting data valuation methodologies relies on the capacity of the target audience to grasp the intended message. Practitioners insist on the power of impact-based data valuation methods, able to convey value by creating data narratives. Data wrapped in stories are 22 times more memorable than bare facts[58] and this is where cities can tap into their communication capabilities, by making citizens understand how the data that they contribute has the capacity to drive change in their communities.

[58] http://chicagoanalyticsgroup.com/blog/archives/01-2017

15

Does Everything Conform to Legal, Ethical, and Data Protection Principles?

Marina Da Bormida in Cugurra

Genoa, Italy
Email: marina.cugurra@eta-one.com

Abstract

In this chapter, the legal and ethical sources relevant to personal data sharing systems, both already applicable and under development, are identified and the main challenges related to personal data sharing operations and practices are sketched, as well as the technology-related opportunities to tackle them, with special attention to what DataVaults[1] technological fixes are going to offer to the data marketplace in this regard towards a trustworthy personal data sharing environment. The chapter also offers an overview of the ethics-related work performed, tools employed, and achievements reached in the DataVaults project: such tools and work could be replicated in other environments, with some adaptations, with the same goal of ensuring the adherence to the relevant legislation, especially GDPR, and to the ethical mandates.

15.1 Introduction

This chapter provides an overview of the legal and ethical framework relevant to personal data sharing systems, focusing on both the existing regulatory sources and on the reforms under development. Such reforms are expected to shape, among other, the future personal data economy, seeking to address the main challenges and barriers related to the operations and practices rotating around personal data. The chapter moves on with a deep-dive on some of

[1] https://www.datavaults.eu/

DOI: 10.1201/9781003399384-16

such challenges and some examples of technology-related opportunities to tackle them with under the current and future European regulatory regime, in particular, underlying what DataVaults technological fixes are going to offer to the private and urban-scale data sharing platforms operating with personal data. Besides these insights on how DataVaults platform in its whole or some of its privacy-preserving technological artefacts could support a trustworthy personal data sharing environment for the benefit of all the involved stakeholders, the chapter also offers an overview of the ethics-related work, tools, and achievements characterising DataVaults itself, from its ethical policy and legal and ethical requirements elicitation, to the ethics and data protection impact assessment methodology used in its piloting activities to assess the legal compliance and ethical soundness of the project's technologies and their use in real-life contexts. These activities and tools can be used in other environments, with some adaptations, with the same goal of ensuring the adherence to the relevant legislation, especially GDPR, and ethical mandates. The chapter ends by drawing conclusions.

This chapter and its findings are mainly based on the legal and ethical surveys conducted within the DataVaults project and take inspiration and extracts from it, combined with insights coming from the recent debates and the literature.[2]

15.2 The Evolving Regulatory Framework Relevant to the Personal Data Sharing Platforms

One of the main barriers to the development and growth of the data economy in relation to personal data is the lack of trusted and secure personal data platforms capable of handing back control over the use of personal data to individuals. This shortcoming hampers personal data sharing practices, despite the wide individuals' willingness to share personal data in return for actual benefits, non-necessarily financial.

There is the need for trusted, secure, and value generating data management and sharing platforms for personal data, allowing stakeholders' collaboration in order to support their own goals and operations, as well

[2] DataVaults Consortium, D2.1 "Security, Privacy and GDPR Compliance for Personal Data Management" (2020).
DataVaults Consortium, D2.3 "Updated DataVaults Security Methods and Market Design" (2021)
DataVaults Consortium, D1.3 "DataVaults MVP and Usage Scenarios", (2021). More information on the DataVaults projects can be retrieved at. https://www.datavaults.eu/

as allowing further stakeholders, such as local communities and local authorities, to offer new socially and environmentally sustainable solutions and business models.

In other words, there is the need for solutions moving forward towards regaining the trust of individuals when it comes to data sharing, letting the control in the hands of the data owners (the individuals) who will be able to decide how, how much, and in which manner they would like to share their personal information, while, at the same time, guaranteeing their privacy and adequate security levels as well as ensuring fair share of the value that their data generate, also in case of secondary use.

This approach is aligned with the European Commission's vision of personal data sharing that should provide benefits for all the actors in the value chain.

Both at European level and in the society, it is emerging the perception that personal data spaces should be promoted, especially on a EU-wide level, ensuring the legal compliance and fostering trust and collaboration. This vision includes addressing the concerns on security, privacy, ethics, and IPR ownership for prioritising human wellbeing and fundamental rights in the data-driven economy.

The recent works "A European Strategy for Data"[3] and the "White Paper on Artificial Intelligence",[4] which represent two pillars of the new digital strategy of the Commission, underline this vision for putting people first in developing technology, defending and promoting European values and rights in any design, development, and deployment of the technology in the real economy.

Any personal data sharing platform should fully embrace this strategy and the promotion of such values, including protection of privacy. In order to do so and foster the creation of a single market for data upholding such values and fully respecting individuals' rights and freedoms, the compliance with the existing legal sources is first of all paramount, and also the adequate consideration of the regulatory reforms under development.

[3] COM/2020/66 final, Communication From The Commission To The European Parliament, The Council, The European Economic And Social Committee And The Committee Of The Regions. "A European strategy for data". Available at: https://eur-lex.europa.eu/legal-content/EN/TXT/?uri=CELEX%3A52020DC0066

[4] COM/2020/65 final, "WHITE PAPER On Artificial Intelligence - A European approach to excellence and trust". Available at: https://eur-lex.europa.eu/legal-content/EN/TXT/?uri=CELEX%3A52020DC0065&WT_mc_id=Twitter

15.3 Existing Regulatory Framework

The consideration of the overall regulatory and ethical framework relevant to the personal data sharing, comprising a number of applicable instruments to be addressed in a systematic way, is key to design, develop, deliver, and operate the personal data platforms in an ethical, private, and fairness-friendly way, which is at the same time compliant with the legislation, and where individuals are enabled to take ownership and control of their data and share them at will, while value is properly attributed to all the entities involved in generating the same. This section does not present a comprehensive analysis of the European regulatory framework, which would fall outside the scope of this work. On the contrary, it indicates the main instruments that are functional to the objective mentioned above.

GDPR, "General Regulation on data protection". The first piece of legislation to mention is the GDPR, "general regulation on data protection 2016/679, of the European Parliament and of the Council of 27 April 2016 on the protection of natural persons with regard to the processing of personal data and on the free movement of such data".[5] It repealed the Directive 95/46/EC (General Data Protection Regulation), providing a comprehensive reform of data protection rules in the EU, establishing common European rules to ensure that personal data enjoys a high standard of protection everywhere in the EU.

One of the main objectives of the GDPR is to give back individuals the control over their personal data, thereby acting as key enabler of the Digital Single Market: personal data can only be gathered and handled legally under strict conditions and for a legitimate purpose. The individuals or organisations collecting or managing personal information have to protect it from misuse and have to respect the data subject's rights, whilst the data subject is enabled to complain and obtain redress if his/her data is misused.

The whole legal source might be relevant to the sharing of personal data within a data platform.

Directive 2002/58/EC "ePrivacy Directive". Another instrument relevant to a personal data sharing platform is the "ePrivacy Directive" (Directive 2002/58/EC on privacy and electronic communications[6]), which replaced the

[5] European Commission, "General Regulation on data protection 2016/679, of the European Parliament and of the Council of 27 April 2016 on the protection of natural persons with regard to the processing of personal data and on the free movement of such data". Available at: https://eur-lex.europa.eu/eli/reg/2016/679/oj

[6] Directive 2002/58/EC of the European Parliament and of the Council of 12 July 2002 concerning the processing of personal data and the protection of privacy in the electronic communications sector (Directive on privacy and electronic communications). Available at: https://eur-lex.europa.eu/legal-content/EN/TXT/HTML/?uri=LEGISSUM:l24120&from=EN

Directive 97/66/EC and was partially amended by Directive 2009/136/EC. It pertains to the processing of personal data and the protection of privacy in the sector of electronic communications, telecommunications networks, and internet services, transposing in the telecommunications sector, which is a "sensitive" area from a privacy perspective, the main principles and rules of the GDPR, aiming at particularising and complementing the former (for instance, as regards the consent to the use of cookies and opt-outs) in case electronic communications data are personal data. Several provisions might be relevant in relation to the exchange of personal data, such as Article 2, on the traffic data and location data, Article 4, on the obligation of adopting security measures appropriated to the risk presented, Article 5, dwelling on the protection to confidentiality of the communications among individuals, Article 6, on user's consent, Article 15, on data retention, and others. The ePrivacy Directive is expected to be repealed by the ePrivacy Regulation.

Human rights law: This area of law includes, among other sources, above all, the European Convention on Human Rights[7] and the Charter of Fundamental Rights of the European Union.[8] Both of them acknowledge privacy and data protection as fundamental human rights in Europe.

From an international perspective, the Universal Declaration of Human Rights (1948) is also relevant: it recognises the privacy as a fundamental human right by protecting territorial and communications privacy. Its Article 8 deals with private and family life, home, and correspondence of the citizen. Since then, more enforceable European tools surpassed its application in the field of data privacy. Article 8.2 states the lawfulness criterion, in the meaning of rule of law.

The European Court of Human Rights' jurisprudence has to be taken into account in relation to personal data sharing practices and tools. This case law is an essential factor supporting the application of these legal sources in relation to the technological artefacts supporting the personal data sharing.

Ethics and soft law instruments. The composite regulatory system to be taken into account also comprises the soft law sources (quasi-legal instruments), which may not have any legally binding force but is helpful in so far they serve to fill in gaps, identify safeguards, boundaries, and obligations to ensure the legitimacy and fairness of personal data sharing platforms, and,

[7] The European Convention on Human Rights, adopted in 1950 and entered into force in 1953. The Convention and its Protocols can be retrieved at the following link:https://www.coe.int/en/web/conventions/search-on-treaties/-/conventions/treaty/results/subject/3

[8] Charter of Fundamental Rights of the European Union, 2016/C 202/02. It can be retrieved at the following link: https://eur-lex.europa.eu/legal-content/EN/TXT/HTML/?uri=CELEX:12016P/TXT&from=EN

at the same time, contributed to find out, on a case-by-case basis, a balance between competing interests. Soft law has an array of possible benefits and usually runs within the boundaries set by its interplay with the traditional legal instruments, in a landscape of increasingly dynamic cross-fertilisation of regulations and technology. It should receive the appropriate consideration when determining personal data sharing technology design and deployment, especially due to the rapidly developing field of data sharing ecosystems: thanks to its flexible nature, that lets it be quickly adapted to future technological progress, soft law could provide useful insights, recommendations, and indications and support in identifying the adequate safeguards and mechanisms in relation to transparency and accountability.

Among the other soft law sources, we can mention, for instance, the European Commission's Communications "AI for Europe" (25 April 2018) and "Building Trust in Human-Centric AI" (8 April 2019), as well as the "Data Protection in the era of Artificial Intelligence. Trends, existing solutions and recommendations for privacy-preserving technologies" (BDVA,[9] October 2019) and "Meeting the challenge of Big Data. A call for transparency, user control, data protection by design and accountability" (Opinion 7/2015, European Data Protection Supervisor, 2015).

Regulation 910/2014 on electronic identification and trust services for electronic transactions in the internal market (eIDAS).[10] This source, repealing the Directive 1999/93/EC, is potentially relevant to the personal data sharing platform. It is aimed at ensuring the proper functioning of the internal market, facilitating seamless digital transactions among individuals and businesses across the same, and creating a climate of trust in online and digital transactions. According to Article 2, it applies to electronic identification schemes notified by a Member State and to trust service providers established in the Union.

This regulation consists of two main parts: one concerns electronic identification, whilst the other concerns trust services (electronic signatures and other trust services).

It sets the conditions for the recognition of electronic identification means of natural and legal persons, the rules for trust services (especially for electronic transactions), besides introducing a legal framework for electronic signatures, electronic seals, electronic time stamps, electronic documents,

[9] Big Data Value Association

[10] Directive 1999/93/EC of the European Parliament and of the Council of 13 December 1999 on a Community framework for electronic signatures. Available at: https://eur-lex.europa.eu/legal-content/EN/TXT/?uri=celex%3A31999L0093

electronic registered delivery services, and certificate services for website authentication. Its provisions regarding the electronic registered delivery services might be relevant to the personal data sharing platforms, since their services can fall into such concept. In fact, the electronic registered delivery service is defined by eIDAS a "service that makes it possible to transmit data between third parties by electronic means and provides evidence relating to the handling of the transmitted data, including proof of sending and receiving the data, and which protects transmitted data against the risk of loss, theft, damage or any unauthorised alterations" (Art. 3, (36) eIDAS). On the other hand, Article 2 (2) eIDAS states that this regulatory source does not apply to "the provision of trust services that are used exclusively within closed systems resulting from national law or from agreements between a defined set of participants". In any case, the eIDAS Regulation states that the processing of personal data must be carried out in accordance with the GDPR and respecting its principle of confidentiality and security of processing.

Regulation 2018/1807 on a framework for the free flow of non-personal data in the European Union, adopted by the EC, applies to any form of data other than personal data, as defined in Article 4.1 of the GDPR. It is functional to create a comprehensive and coherent approach to the free movement and portability of data in the EU. Notably, its main objectives are to further promote the free movement of data and data processing services (Recital 4), whilst facilitating cross-border availability of data, enhancing legal certainty and creating a level playing field through a single set of rules for all market participants. It supplements and complements the GDPR in issues related to non-personal data within the Digital Single Market, primarily concerning business and public sector users of data storage and processing services. This instrument should be taken into account in relation to the non-personal data (such as insights, other derivatives, data related to the persona, and data completely anonymised) collected, shared, and used in the personal data platform, as well as for mix platform, combining personal and non-personal data sharing.

E-Commerce Directive, Directive 2000/31/EC on certain legal aspects of information society services, in particular electronic commerce, in the internal market. This is another important legislative source that might be relevant for the operation of personal data sharing platform, considering that their services, to the extent that they represent information society services, might be provided for remuneration, at a distance, by electronic means and at the individual request of a recipient of the service and, therefore, fall under the scope of this Directive. Considering the nature of this source, the national provisions implementing it would need to be considered in each country where the personal data platform is adopted. Section 4 on intermediary liability may be particularly relevant in the case of illicit third-party content.

Platform-to-Business Regulation – P2BR. The Regulation 2019/1150 on promoting fairness and transparency for business users of online intermediation services is a set of rules in the area of business platforms for creating a fair, transparent, and predictable business environment for smaller businesses and traders on online platforms, in order to enable consumers to receive the highest quality goods and services. The P2BR, which is part of the legislative measures promoted by the EC for the Digital Single Market strategy, foresees a list of measures ensuring transparency and fairness with the intent to temper the natural asymmetries that characterise the relationship between the platforms and their suppliers, establishing a fair and trustworthy innovation-driven ecosystem. Its Article 2 describes a set of requirements of the intermediation services (platforms) that fall into the scope of its application. Its definition of intermediaries describes only the services that have a direct relationship with business users and their clients without a clear threshold, applying indistinctively to all types of platforms falling in such criteria. The two main principles set by the P2BR are transparency and fairness. Taking into account who the data platform concerned intends to offer its services to, it could fall within the P2BR scope. Nevertheless, in order to be applicable to such a platform, it should fall under the concept of online intermediation service: whilst it is likely that the data providers are businesses, it is not sure that the data receivers are consumers, as requested by the definition of the online intermediation service, which is, in principle, applicable only for business users.[11]

Directive 2019/770 on certain aspects concerning contracts for the supply of digital content and digital services might also be relevant. Given that contracts are often crucial for the personal data platform, it is paramount to consider the EU framework related to contractual agreements, which may be applicable in the context of the project. From a consumer policy perspective, considering the steps taken by the EC to implement a "digital update" of consumer contract law, it is widely recognised that consumers should enjoy the same level of protection under consumer contract law, whatever the object of consumption is. This Directive aims at the maximum harmonisation

[11] Such services must have the following characteristics: being information society services, i) allowing business users to offer goods or services to consumers for facilitating the initiating of direct transactions between such business users and consumers ii) and provided to business users on the basis of contractual relationships between the provider of those services and business users (which, in turn, offer goods or services to consumers).

and at introducing mandatory contractual liability for the non-conformity of digital content with the contract. It also extends the information duties as well as the right to withdraw from a contract in case of "free digital services" contracts, where consumers provide personal data instead of paying a fee. The Directive is directed to protect the consumer, understood as "any natural person who, in relation to contracts covered by this Directive, is acting for purposes which are outside that person's trade, business, craft, or profession" (Article 2.6).[12] The Directive applies to "contracts of an indefinite or fixed duration which were concluded before the application date and provide for the supply of digital content or digital services over a period of time, either continuously or through a series of individual acts of supply, but only as regards digital content or a digital service that is supplied from the date of application of the national transposition measures", with the exception of the provisions on the modification of the digital content or digital service and the right to redress. In relation to contractual agreements and consumer protection, also the following pieces of legislation can be considered: Directive 93/13/EEC on unfair terms in consumer contracts and Directive 2019/2161 (amending Council Directive 93/13/EEC and Directives 98/6/EC, 2005/29/EC, and 2011/83/EU) as regards the better enforcement and modernisation of Union consumer protection rules.

Security Law. Despite from a legal point of view the requirements related to security are mainly coming from the GDPR and the ePD, it is useful to take into account the latest legislative developments in this area. Cybersecurity has been identified as one of the highest priorities for the EU: the achievement of a secure and safe environment is a precondition to enhance trust and to boost business opportunities. In this area of law, it is important to mention the following.

- The Directive 2016/1148 on security of network and information systems (NIS), which was part of the 2013 EU cybersecurity strategy, comprising binding and non-binding legal instruments aimed at establishing a high standard of security across the European Union. It applies to operators of essential services and digital service providers.
- The Regulation (EU) 2019/881 (Cybersecurity Act), included in the Cybersecurity Package. It provides rules on the creation of an EU cybersecurity certification scheme for ICT products, ICT services, and ICT

[12] Member states can extend the protection afforded to other persons who are not qualified as consumers.

processes and aims to improve the cross-border coordination, besides promoting EU standards. The cybersecurity certification schemes for ICT products, ICT services, and ICT process might be of interest for a personal data platform, since it can enhance its security and trust.

As regards the EU encryption framework, the following documents are particularly interesting for the personal data market: the ENISA[13] Opinion Paper on encryption (2016) and the European Electronic Communications Code (EECC), established with the Directive 2018/1972. This code, in its security provisions, makes reference to encryption protocols and, explicitly, to the end-to-end encryption.

15.4 The Regulatory Reforms Under Development

Vast reforms are underway and an update of the European regulatory landscape was announced in terms of the Commission's Mission Statement for 2019–2025. Especially some of them are expected to be significant for the deployment and use of personal data platform.

Considering the envisaged role of individuals as data owner, it is opportune to follow the developments in terms of the European consumer protection framework and, more specifically, the developments related to so-called "New Deal for Consumers" initiative, adopted in 2018.[14] This initiative is functional to achieve a stronger and better enforced consumer protection rules in light of a growing risk of EU-wide infringements and at modernising EU consumer protection rules in view of market developments.

Another important regulatory source is the Communication "2030 Digital Compass: the European way for the Digital Decade".[15] Its Vision for 2030 relies on empowered citizens and businesses. The Communication also underlines the need to full respect of EU fundamental rights, including the freedom of expression (including access to diverse, trustworthy, and transparent information), the freedom to set up and conduct a business online, the protection of personal data and privacy and right to be forgotten, and the

[13] European Union Agency for Network and Information Security

[14] Communication of the Commission of 11 April 2018—A New Deal for Consumers, (COM)2018, 183 final. Available at: https://eur-lex.europa.eu/legal-content/EN/TXT/?uri=CELEX%3A52018DC0183

[15] COM(2021) 118 final. Communication from the Commission to the European Parliament, the Council, the European Economic and Social Committee and the Committee of the Regions, "2030 Digital Compass: the European way for the Digital Decade". Available at: https://eur-lex.europa.eu/legal-content/en/TXT/?uri=CELEX%3A52021DC0118

protection of the intellectual creation of individuals in the online space. It is envisaged the definition of a comprehensive set of digital principles allowing to inform users (besides guiding policy makers and digital operators), including, for instance, a secure and trusted online environment, the access to digital systems, and devices that respect the environment, accessible and human-centric digital public services and administration, ethical principles for human-centric algorithms, and access to digital health services. The EC proposed to include these sets of digital principles and rights within an inter-institutional solemn declaration between the European Commission, the European Parliament, and the Council.

The overarching regulatory framework especially relies on the already mentioned new European Data Strategy. It was presented along with the Commission's Communication on "Shaping Europe's digital future": data are embraced as the "lifeblood of economic development". Therefore, the EC aims at renewing its overarching framework to achieve the proper balance between, on the one hand, the wide availability and use of data and, on the other hand, the high preservation of privacy, security, safety, and ethical standards. Aspects related to data ownership and data governance are going to be addressed and/or reframed. The Strategy is motivated by the need to put people first in developing technology and to defend and promote European values and rights in how the technology is designed and deployed in the real economy. The Strategy sets out a programme of policy reforms, already started with the Data Governance Act, the Digital Services Act, the Digital Markets Act, and the Cybersecurity Strategy.

The proposal **Data Governance Act** (DGA) was published in November 25, 2020 and has been conceived to play a vital role in ensuring the EU's leadership in the global data economy, whilst empowering users to stay in control of their data. The DGA sets out policy measures and investments designed to capitalise on European vast quantity of data and, hence, to give the EU businesses a competitive advantage. The envisioned framework is expected to boost data sharing, encouraging a greater reuse of data by increasing trust in data intermediaries and strengthening various data-sharing mechanisms across the EU. In addition, the DGA will support the creation of EU-wide common, interoperable data spaces in strategic sectors relevant to the personal data platform, such as health, energy, and mobility, which, in turn, are meant to bring benefits to citizens. Its broad definition of data includes personal data as defined in the GDPR, which apply simultaneously to the DGA. As remarked by the explanatory memorandum, the DGA and its measures are fully compliant with the data protection legislation and increase, in practice, the control that individuals have over the data that they generate.

This is an important element for the personal data economy and the personal data platforms in particular. Many of its rules are potentially relevant for the private and urban personal data platforms. They include, among others:

- conditions for reuse of public sector data, which are subject to existing protections (such as intellectual property, commercial confidentiality, and data protection);
- obligations on providers of various types of intermediation services within data-sharing services – new European rules on neutrality are defined to allow novel data intermediaries to function as trustworthy organisers of data sharing;
- a set of measures to increase trust in data-sharing, due to the fact that the lack of trust is currently a major obstacle and results in high costs;
- data altruism, providing its concept and the possibility for organisations to register as "Data Altruism Organization recognized in the Union";
- measures to give the individuals the control on the use of the data they generate, in particular by making it easier and safer for companies and natural persons to voluntarily make their data available for the wider common good under clear conditions.

On the other hand, the proposal European Digital Service Act (DSA)[16] is expected to update and reform the framework established by the e-Commerce Directive, addressing the topics of intermediary liability and safety rules for digital platforms, including transparency, information obligations, and accountability for digital services providers. At the same time, there is a strong call for maintaining the core principles of the e-Commerce Directive, its measures having the consumer protection at their core and the protection of fundamental rights in the online environment, as well as online anonymity wherever technically possible. In fact, the DSA builds on the key principles set out in the e-Commerce Directive, which is still applicable, seeking to ensure the best conditions for the provision of innovative digital services in the internal market, to contribute to online safety and the protection of fundamental rights, whilst setting a robust and durable

[16] COM/2020/825 final, Proposal for a REGULATION OF THE EUROPEAN PARLIAMENT AND OF THE COUNCIL on a Single Market For Digital Services (Digital Services Act) and amending Directive 2000/31/EC. Available at: https://eur-lex.europa.eu/legal-content/en/TXT/?uri=COM%3A2020%3A825%3AFIN

governance structure for the monitoring and supervision of providers of intermediary services.

Since the adoption of the e-Commerce Directive, novel information society (digital) services have emerged, which, on the one hand, have deeply contributed to societal and economic transformations in the European Union and worldwide but, on the other hand, have brought new risks and challenges, both for society as a whole and for individuals using such services. The DSA, which is envisaged to be a standard-setter at global level, addresses the online marketplaces and consumer trust in the digital economy, while respecting users' fundamental rights and advocating for rules to underpin a competitive digital environment in Europe. Clear responsibilities and accountability are defined for providers of intermediary services, and in particular online platforms, including marketplaces. Due-diligence obligations are set for certain intermediary services in order to improve users' safety online across the entire Union and improve the protection of their fundamental rights. Certain online platforms have the obligation to receive, store, partially verify, and publish information on traders using their services in order to ensure a safer and more transparent online environment for consumers. A higher standard of transparency and accountability is set for certain platform as well as obligations to assess the risks their systems pose and to develop appropriate risk management tools to protect the integrity of their services against the use of manipulative techniques. However, the operational threshold for service providers in scope of these obligations includes only online platforms with a significant reach in the European market (currently set to more than 45 million recipients of the service). The DSA is without prejudice to the GDPR.

The proposal Digital Market Act[17] might be relevant to the personal data architectures in the future. Its objective is "to allow platforms to unlock their full potential by addressing at EU level the most salient incidences of unfair practices and weak contestability" in view of allowing end-users and business users alike to reap the full benefits of the platform economy and the digital economy at large, in a contestable and fair environment. Nevertheless, its scope of application concerns "markets characterised by large platforms, with significant network effects acting as gatekeepers".

The **proposal of Regulation on Privacy and Electronic Communications (ePrivacy Regulation)** is another legal instrument under development to

[17] COM/2020/842 final, Proposal for a Regulation of the European Parliament and the Council on contestable and fair markets in the digital sector (Digital Markets Act). Available at: https://eur-lex.europa.eu/legal-content/en/TXT/?uri=COM%3A2020%3A842%3AFIN

monitor is the ePrivacy Regulation,[18] intended to update European privacy framework, repealing the ePrivacy Directive, for a better alignment of the provisions of such Directive with those of the GDPR, while addressing the new challenges to privacy, brought about by the significant advancement of technology the last two decades. In fact, albeit objectives and principles of the existing framework remaining sound and relevant, the essential technological, economic, and business progresses, together with the ever-increasing penetration of the internet in various aspects of the life and its vital role in the Digital Single Market, call for the modernisation of the Directive. The choice of a Regulation is meant to improve the harmonisation. As clarified in the proposal itself, it will be "*lex specialis*" to the GDPR: it will fine-tune and complement the GDPR as regards electronic communications data that qualify as personal data, whilst all matters concerning the processing of personal data not covered by the proposal remain regulated by the GDPR.

The **proposal for a Directive on measures for a high common level of cybersecurity across the Union**[19] pertains to the area of security and will repeal the Directive (EU) 2016/1148. This proposal is directed to introduce systemic and structural changes to the current NIS Directive for covering a wider set of entities across the Union, with stronger security measures, such as mandatory risk management, minimum standards, and relevant supervision and enforcement provisions. As highlighted by the European Data Protection Supervisor,[20] it is essential to integrate "the privacy and data protection perspective in the cybersecurity measures stemming from the Proposal or from other cybersecurity initiatives of the Strategy in order to ensure a holistic approach and enable synergies when managing cybersecurity and protecting the personal information they process", and that "all cybersecurity systems and services involved in the prevention, detection, and response to cyber threats should be compliant with the current privacy and data protection framework".

---

[18] COM/2017/010 final, Proposal for a Regulation of the European Parliament and the Council concerning the respect for private life and the protection of personal data in electronic communications and repealing Directive 2002/58/EC (Regulation on Privacy and Electronic Communications). Available at: https://eur-lex.europa.eu/legal-content/EN/TXT/?uri=CELEX%3A52017PC0010

[19] COM/2020/823 final, Proposal for a Directive on measures for a high common level of cybersecurity across the Union , repealing Directive (EU) 2016/1148. Available at: https://eur-lex.europa.eu/legal-content/EN/TXT/?uri=COM%3A2020%3A823%3AFIN

[20] European Data Protection Supervisor, "Opinion 5/2021 on the Cybersecurity Strategy and the NIS 2.0 Directive", 2021. Available at: https://edps.europa.eu/data-protection/our-work/publications/opinions/edps-opinion-cybersecurity-strategy-and-nis-20_en (accessed Jul. 29, 2022)

In parallel, the EC and the High Representative of the Union for Foreign Affairs and Security Policy issued a Joint Communication titled "The EU's Cybersecurity Strategy for the Digital Decade", whose overall objective is to ensure a global and open internet with strong safeguards for the risks to security and the fundamental rights, in a multi-stakeholder model.

15.5 Main Legal and Ethical Challenges and Technology-enabled Opportunities to Tackle with Them

The operation of a personal data platform might imply a number of legal and ethical challenges, for instance, related to personal data management in terms of data collection, data sharing and processing, as well as to the potential trade-off between the need to maximise data utility whilst protecting human rights and preserving meaningful human control or the question if and to what extent the future technological development should allow for automation of (legal) protection in an increasingly digital society.[21]

In the following paragraphs, some important challenges and trends related to the tools and technologies aimed at facilitating secure and trustworthy data sharing in an urban and private environment are provided, taking inspiration and extracts from the work and regulatory surveys conducted within the DataVaults project,[22] also in this case combined with insights from recent debates and the literature.

15.6 The Need to Avoid Consent Fatigue and to Develop and Use User- and Data-Protection-Friendly User Interface

According to the GDPR, the consent has to be given for the processing of personal data for one or more specific purposes. In case of new purposes, it is necessary to either get fresh consent specifically covering such new purpose or find a different legal basis for the new purpose.

---

[21] Big Data Value Association (BDVA), "Data protection in the era of Artificial Intelligence", 2019. Available at: https://www.bdva.eu/data-protection-era-artificial-intelligence-0 (accessed Jul. 29, 2022)

[22] DataVaults Consortium, D2.1 "Security, Privacy and GDPR Compliance for Personal Data Management", 2020 DataVaults Consortium, D2.3 "Updated DataVaults Security Methods and Market Design", 2021 DataVaults Consortium, D1.3 "DataVaults MVP and Usage Scenarios", 2021

Even when expressed through electronic means, the consent of the data subject should be preventive and unambiguous. It requires a statement or clear affirmative action of the data subject. For instance, these actions can consist of ticking a box in an online environment, the choice of technical settings for information society services, and any other statement or conduct clearly indicating the data subject's acceptance of the data processing activities.

In a personal data sharing platform, it is also necessary to ensure that, where consent is obtained through the use of a service-specific user interface (for example, within a given personal data app or the interface of an IoT device), the individual must be able to withdraw consent through the same electronic interface with undue effort and without detriment.

The EDPS Opinion 7/2015[23] outlines challenges relevant to data platforms entailing the sharing of personal data and that need to be addressed. It clarifies that in many big data environments "individuals cannot efficiently exercise control over their data and provide meaningful consent in cases where such consent is required. This is all the more so as the precise future purposes of any secondary use of the data may not be known when data is obtained: in this situation, controllers may be unable or reluctant to tell individuals what is likely to happen to their data and to obtain their consent when required".

The data collection and processing in such data platforms might be intended for multiple purposes and it is necessary to ensure the consent for all of these purposes (Recital 32 GDPR).

Recital 43 GDPR casts doubt on an approach based on one single consent form, broadly formulated as pre-emptively covering different future business models of the data controller.

Globalised, generic consent for multiple vague purposes risk to be assumed as not freely given and the question that arises is whether separate consent and the need for several, broken down consent requests are appropriate. This needs to be explored in the context of DataVaults, but also reflecting on the need to avoid "consent-fatigue" of a data subject.

As acknowledged by the Article 29 Working Party, a layered approach could be a possible solution, still providing all necessary information step by step and providing balancing means of user control, whilst being substantially different by the mere use of pre-ticked boxes: it is not necessary that the first layer of information is completely in-depth about the details of the processing. It should be explored if, for most of the cases (though not

[23] https://edps.europa.eu/sites/edp/files/publication/15-11-19_big_data_en.pdf

applicable to the special categories of personal data of Article 9 GDPR), an implicit consent (such as a shade going away after a few seconds and assumes "yes") could work, after the first general consent during the installation of the service. It should be likewise investigating which information needs to be given to the data subject in which layer.

Useful indications can be retrieved in the following GDPR Recitals:

1. Recital 32, which clarifies that it can be a written statement, including by electronic means, or an oral statement, if the data subject's behaviour clearly indicates his/her acceptance of the data processing. It is recommended that if the data subject's consent is to be given following a request by electronic means, such a request must be clear, concise, and not unnecessarily disruptive to the use of the service for which it is provided.
2. Recital 33, which states that, being often not possible to fully identify the purpose of personal data processing for scientific research purposes at the time of data collection, data subjects should be allowed to give their consent to certain areas of scientific research (or parts of research projects) when in keeping with recognised ethical standards for scientific research. "Data subjects should have the opportunity to give their consent only to certain areas of research or parts of research projects to the extent allowed by the intended purpose".
3. Recital 42, which states that "...For consent to be informed, the data subject should be aware at least of the identity of the controller and the purposes of the processing for which the personal data are intended. Consent should not be regarded as freely given if the data subject has no genuine or free choice or is unable to refuse or withdraw consent without detriment".

The data platform consent management policies need to ensure that the consent is:

1. "granular", capable of providing distinct consent options for distinct processing operations;
2. specific to "one or more specific" purposes, ensuring that the data subject has a choice in relation to each of them;
3. freely given, in the sense that the data subject should be able to exercise a real choice, without risk of deception, coercion, intimidation, or significant negative consequences if he/she does not consent;

4. informed, being the provision of information to data subjects prior to obtaining their consent necessary to enable them to understand what they are agreeing to, make informed decisions, and exercise control, and, in general, their rights (including to withdraw their consent). As noted, a layered approach could help in this regard;

5. separate from other terms and conditions;

6. "explicit", in case of processing of special categories of data, profiling activities or cross-border data transfers. Though in many cases, the term "explicit" could be interpreted as given in writing with a hand-written signature, in digital or online context like DataVaults, a data subject may be able to issue the required statement with other modalities (such as by filling in an electronic form, or by using an electronic signature).

These considerations are especially applicable to private data platforms, whilst for urban data platform, the personal data collection and/or use might also relies on other legitimate sources of the processing pursuant to Article 6 GDPR "Lawfulness of processing", in particular points: "(e) processing is necessary for the performance of a task carried out in the public interest or in the exercise of official authority vested in the controller; (f) processing is necessary for the purposes of the legitimate interests pursued by the controller or by a third party, except where such interests are overridden by the interests or fundamental rights and freedoms of the data subject which require protection of personal data, in particular where the data subject is a child".

In case of public authorities, there might be a clear imbalance of power in the relationship between the controller and the data subject and other lawful bases for the processing could be, in principle, more appropriate. This has to be taken into consideration for the urban data platform relying on personal data use. Pursuant to the accountability principle, the existence of valid consent must be demonstrable by the data controller (accountability).

In strict correlation with this topic, a personal data platform or application also requires to adopt user and data protection friendly user interface (UI), capable of facilitating as much as possible the user control features and consent management in an easy manner. It should be capable of collecting consent and constraints/restrictions, providing appropriate options for user information and control, thereby enabling the data subject to easily consent and exercise his/her rights set forth under data protection legislation, at national and European level.

An important element to consider is the wide range of data sources and to pay special attention in case where it includes sensitive information in the sense of Article 9 GDPR.

A filter on those data categories could allow the UI to distinguish between consent requests on "normal" personal data and those involving sensitive data. It could be investigating whether introducing functionalities for automatically detecting when sensitive data (or particular subset of sensitive data, for instance, in the healthcare demonstrator) is collected, using machine learning techniques or other techniques and filtering such data.

The following challenges could occur and need to be addressed:

- managing consent in a fine-grained way (including, for instance, partial granting or withdrawal of consent in some circumstances);
- managing the own data and exercise data subject's rights in an easy way, for instance, as regards adding, deleting, and rectifying personal data, and including also the possibility to access additional information in case of a data breach;
- switching back and forth between different consent modalities, such as always requiring explicit consent for personal data sharing in some situations and opting for convenient assumption of implicit consent in other;
- ensuring data portability and exporting the own personal information (for instance, in an RDF format).

15.7 Risk-based Approach and Risk-Exposure Dashboard

Within a data sharing ecosystem, it is advisable, in relation to ethics risks and especially to those related to personal data collection and/or processing, to adopt a risk-based approach, following the current regulatory trend, as provided, for instance, by the GDPR (Recitals 75 and 76) and AI Act proposal.

This approach requires to consider the risk of varying likelihood and severity for the rights and freedoms of natural persons. Following this approach, it is therefore necessary to evaluate the ethics risks related to the data processing activities of the platform, assessing the particular likelihood and severity of each risk to data protection (or other ethical values), taking into account "the nature, scope, context and purposes of the processing and the sources of the risk". The assessment of the risk must be conducted in an objective manner to determine whether there is a "risk" or a "high risk", in order to let the data controller be particularly prudent to carefully consider their obligations when necessary. Such an approach requires consideration of what measures are appropriate in each case, depending on the scope, nature, context, and purposes of the processing concerned, as well as of the risks of varying likelihood and severity for freedoms and rights of individuals.

The more severe and likely the risks from the proposed processing, the more measures will be required to counteract such risks.

According to Recital 75, examples of potentially risky processing relevant to a platform enabling the exchange of personal data include: i) processing that may give rise to discrimination, identity theft, financial loss, reputational damage, unauthorised reversal of pseudonymisation, or any other significant economic or social disadvantage; ii) processing that might deprive data subjects of their rights and freedoms or prevent them from exercising control over their personal data; iii) processing of sensitive personal data; iv) processing for purposes of profiling; v) processing of personal data of vulnerable natural persons; vi) processing involving a large amount of personal data and affecting a large number of data subjects.

For operationalising the risk-based approach, the DataVaults project developed a specific tool and related service: the Risk Exposure Dashboard, displaying an individual's current and projected risk estimations, which are updated whenever a modification to the shared assets occurs. Such estimations and risk exposure metrics are calculated relying on the data assets the data owner has already shared, as well as on the data they intend to share, and taking into account all sharing aspects, such as anonymisation level and discoverability, as well as the information provided by the nature of the data itself. The calculation of the privacy risk exposure, based on previous knowledge and depending on the data already available and shared and specific metrics, will allow also being able to notify individuals of their privacy risk exposure from the DataVaults Cloud Platform through the DataVaults Personal App.

The Risk Management Service might represent a high-value powerful accountability tool for the fulfilment of the GDPR-compliant informed consent requirement and user control, strengthening the positioning on the market of a (urban and/or private) data platform embedding it within its architecture and offering it to the individuals to foster their inclination to share their personal information.

A dedicated "Sharing Risk Information" operation is, indeed, essential for raising the awareness of the individuals on the privacy exposure impact of sharing data assets.

15.8 Personas and Digital Twins

Depending on the defined data sharing configuration and selected user privacy level, different tools and techniques for privacy enhancement will be used, ranging from the integration of traditional obfuscation schemes such as digital twins and user personas, to the use of trusted computing technologies (i.e., TPMs) as a central element for the provision of privacy-preserving signature schemes based on the use of Direct Anonymous Attestation.

It is interesting here to elaborate on some legal and ethical challenges and opportunities raised by the personas and the digital twins in relation to the personal data platform or, in any case, to urban data platform based on personal data for their operation and service provision.

In DataVaults, the individuals can select the preferred level of anonymisation for the data asset they are going to upload and share in the DataVaults Cloud Platform: their personal data can be shared without applying anonymisation (eponymous) or in anonymised way by implementing the digital twin generator or the persona group generator. More precisely, if the individual selects the anonymised sharing, depending on the individual's preference to upload and share as personal anonymised data (i.e., digital twin) or as grouped anonymised data (i.e., to become part of a persona group).

This is a very useful functionality related to the anonymisation features and level of the personal data sharing that should be available in any personal data platform.

As regards DataVaults, this is provided through the use of the anonymisation bundle, which, as regards personas generation, groups data coming from different individuals and processes them using statistical methods for creating an aggregated representation/model where the individual's data is obfuscated by being included in a large pool of similar data, the so-called persona.

In DataVaults, personas will be partially auto generated and presented to the data scientists prior to his/her analysis, based on certain similar aspects identified by the system (age group, location, interest, compensation requested, etc.). Though the DataVaults Cloud Platform, the data scientist will be provided with an engine for the creation of aggregated profiles composed of data assets from several individuals sharing certain similarities (generation of these personas).

The creation of such personas is based on the obfuscation and merging of data originating from multiple users with similar characteristics; therefore, it is paramount to preserve their privacy. Such personas are exactly aimed at preserving the privacy and anonymity of the indistinct individuals considered for the specific representation/model though, at the same time, they provide valuable information to data seekers. In DataVaults, it is up to the individual to decide whether to share personal data in this way: the individuals have indicated in the sharing configuration their intention to share data for the personas generation. In any case, their privacy is protected, as all data assets to be shared under this condition, are appropriately anonymised prior to being transferred to the Cloud and being used in one or more personas.

One of the challenges that need further investigation, in this regard, pertains to the revoked consent for data assets used for building personas. All the data processing operations based on consent, which took place before the withdrawal, remain lawful but also that, in principle, any further processing of these data is prevented, if there is no other lawful basis justifying the continued retention and/or processing of the data. It is important to consider whether there is non-expired contract in place comprising such data assets: in that case, it is reasonable to conclude that the withdrawal can be exercised for the future without retroactive effect.

In relation to this issue, it is important to bear in mind different aspects: the individuals' right to withdraw consent anytime, the right to erasure/right to be forgotten and its boundaries (in consideration of the available technology, means, and possible reasonable steps), and the other legitimate grounds for personal data processing and the limits to their applicability, with possible switching from one legal basis to another (for instance, in case of urban data platforms), as well as the interest of the data seekers. The legitimacy and fairness of technologies need to be sought by promoting the balance between competing interests and the determination of the required level of protection for the personal information involved in these cases. However, another concern related to this regards the unlinkability of created user personas, in case of deletion of such selected data assets from any created user personas. In other words, in case some data assets are deleted from the personas, the unlinkability to the user identity from whom (obfuscated) data are also included in these personas should be preserved.

As regards the creation of personas, it has also to be further explored if this implies or not in the specific personal data platform concerned, some "profiling", in the meaning provided by GDPR and therefore whether Article 22 is applicable and, in case it is, if additional measures need to be taken. It needs to be clarified on a case-by-case basis whether the creation of the persona and its use imply or not an automated decision-making. It is important that the human intervention will be part of the task, especially in case some effects on the individuals could occur (such as exclusion/limitation from some service or from a data sharing contract).

On the other hand, when the data scientists create the merged persona, the current user privacy risk exposure, as calculated by the DataVaults Risk Assessment framework, should respect the privacy choices defined by the user (in the data sharing configuration): in other words, the quantified privacy risk exposure values need to be kept within the user acceptable boundaries.

Otherwise, the personal data platform should inform the user of appropriate actions to be taken for privacy enhancement.

Moving to the digital twins, first of all, it is useful to provide a snapshot of the concept. “A digital twin is a digital representation of a physical process, person, place, system or device”. This concept, which emerged in the field of manufacturing domain, refers to digital simulation models that run alongside real-time processes[24] and it is conceptualised as digital replicas of physical entities, made possible by the use of technological breakthroughs as sensing, processing, and data transmission.

The digital twin concept is wide and can cover different aspects in different domains. For instance, there are urban scale digital twins, which are “that are used to simulate environments and develop scenarios in response to policy problems”.[25]

The notion of urban scale digital twin has a central role within the field concerning smart cities design, and although there is not a commonly accepted definition of urban digital twins, the common denominator of the different definitions relies on the “bi-directional mapping relationship that exists between physical space and virtual space” for establishing “real-time connection(s) between the virtual and the real”.[26]

The urban scale digital twins, besides useful for observing, recognising, and understanding the physical world, are also aimed at controlling and transforming it[27] since they entail the capacity to monitor activities in the city but even to use such data captured through monitoring for shaping more efficient and more sustainable cities and services in different areas, such as data concerning traffic and transportation, utilities provisioning, power generation, water supply, and waste management among other.

As acknowledged also by the DUET Project and the Living-in.EU Initiatives, “local digital twins can change the way cities are planned, operated, monitored and managed”.[28]

---

[24] Grieves, M. (2014). Digital twin: manufacturing excellence through virtual factory replication. White paper, 1(2014), 1–7.

[25] Charitonidou, M. (2022). Urban scale digital twins in data-driven society: Challenging digital universalism in urban planning decision-making. International Journal of Architectural Computing, 20(2), 238–253. https://doi.org/10.1177/14780771211070005

[26] Deren, L., Wenbo, Y. & Zhenfeng, S. Smart city based on digital twins. Comput.Urban Sci. 1, 4 (2021). https://doi.org/10.1007/s43762-021-00005-y

[27] Tao, F., & Qi, Q. (2019). Make more digital twins. Nature, 573(7775), 490–491. https://doi.org/10.1038/d41586-019-02849-1

[28] Local Digital Twin - Living in EU. Available at: https://living-in.eu/groups/solutions/local-digital-twin (accessed Jul. 29,2022)

Within this overall debate around the urban scale digital twins and their future potentialities for the research on smart cities and their big data as well as, more in general, within the overall debate on digital twins, this chapter will refer only to the personal digital twins relevant in the framework of a data platform based on personal data and their exchange using the elaboration of the digital replication of individual human data. We can refer to them as personal digital twins, since they reflect an individual (habits, history, behaviour, and social interaction) and their personal data.

In particular, in the DataVaults project, the individual can configure the sharing anonymisation level by selecting the preferred level of anonymisation for the data asset they are going to share, ranging from sharing data without applying anonymisation (eponymous), to anonymise personal data at an individual level (digital twin) or, as already mentioned, to anonymise them at a group level making them available for the creation of personas. In case of selection of this data sharing configuration (digital twins), the DataVaults Cloud Platform shall generate the digital twin of an individual by anonymising and obfuscating personally identifiable data while preserving the valuable information enclosed in the data asset, through the use of the identity provided by another DataVaults component, the Identities Wallet. The individual is allowed to view at any time under which digital twin identities they have shared data anonymously with the DataVaults Cloud Platform. In an urban landscape, the personal data that can, potentially, be part of a personal digital twin, comprise both the small portion of data generated and captured by the individual (self-measurement), as well as, mainly, the data resulting from interaction of the individual with their environment, which will be likely captured by third parties.[29] In an urban ecosystem, the exploitation of the personal digital twins is consistent also with the citizen-centric approach of urban digital twins for the benefit of people themselves by ensuring that people have better experiences in complex situations and will also inform better infrastructure investment decisions.

Some of the ethical challenges potentially raised by the digital twins, for instance, based on data captured through Internet-of-Things-based sensing technologies have been initially explored by the narratives[30] and are illustrated below, though the issues are still open.

[29] Saracco R., Personal Digital Twins: What Data? – IEEE Future Directions, 2022. Available at: https://cmte.ieee.org/futuredirections/2018/01/16/the-rise-of-digital-twins/ (accesses Jul. 29, 2022)

[30] D. Helbing, J.A. Sanchez-Vaquerizo, "Digital twins: Potentials, Limitations and Ethical Challenges", Preprint, 2022.

This regards, for instance, the case of use of personal digital twins to run "smart cities". First of all, privacy and security issues might be entailed due, for instance, to the pervasive mass surveillance implied by ubiquitous measurements, or risk of new kinds of discrimination when the individual's social or medical status is measurable or known and can be used to determine their right to access services, facilities, opportunities, or other.

Of course, from an ethical point of view, there may be further concerns of undesired side effects.

Some of them are related to the nature of the human beings and the fact that people are often complex and adaptive to the changing environment: for instance, people can learn, exchange knowledge, have consciousness, are moved by goals changing over time, have emotions, and so on. Similar characteristics might pose particular challenges for creating digital twins and to tackle with such variables it would be necessary to collect and analyse massive amount of sensitive personal data in order to generate increasingly detailed digital twins and this raises privacy issues and the risk of promotion of a society oriented towards control (dataveillance).

There is the risk that application based on personal digital twins might interfere with individual thoughts, decisions and behaviours, human rights, and human dignity.

Other ethical concerns pertain, for example, on the risk of new forms of identity theft, abuse, and deception and how to mitigate them as well as the risk that people are entirely replaced by digital twins.

Running a city based on personal digital twin could be misused: for instance, by knowing individuals' strengths and weaknesses, there is the risk of tricking or manipulating everybody.[31] Furthermore, "a digital twin of society would also make it possible to determine how much one can pressure people without triggering a revolution, or figure out how to overcome majorities, how to break the will of people, and how to impose policies on them, which do not represent their will",[31] thereby undermining human rights.

A further concern regards the risk that personal digital twins are given greater opportunities and authority than human beings themselves, even though the digital representation of people and their desires could be biased, manipulated, or hacked. A personal digital twin and its data could be given more attention by the system/platform more than to humans, ignoring the opinion of the human the digital twin should represent.

[31] Isaak, J.; Hanna, M.J. User Data Privacy: Facebook, Cambridge Analytica, and Privacy Protection. Computer 51, 8 (2018), 56–59. https://doi.org/10.1109/MC.2018.3191268

There is also the risk of over-simplifications and of neglecting details and human dignity and other hardly measurable aspects, therefore undermining one of the main strengths of social systems: their ability to self-adaptation, self-organisation and co-evolution, or, in other words, of a "technological determinism" of society.

In order to avoid that, people could be managed like things; it is, therefore, paramount in a highly networked, complex urban context characterised by data-driven and AI-empowered solutions, to prevent these risks to materialise by strongly relying on ethical mandates and soft law. We can mention, as examples, the Ethics Guidelines for Trustworthy AI and the UN's "Agenda 2030" with its 17 sustainability development goals (SDGs), as well as UNESCO's Recommendation on the Ethics of Artificial Intelligence,[32] as well as, more in general, the current regulatory reforms under development.

It is also critical that, rather than replacing individual preferences by automated machine decisions, to keep the individual's control on their data and on the decision made relying on them. This is what systems like DataVaults are directed to do, thereby minimising the potential misuse of powerful digital technologies while maximising benefits for the society.

15.9 Challenges Related to Smart Contracts, the eIDAS Regulation, and the Self-Sovereign Identity

In order to set, sustain, and mobilise an ever-growing ecosystem for personal data and insights sharing and to foster an enhanced collaboration between individuals and data seekers capable of rejuvenating the personal data value chain, it is key to secure value flow based on smart contracts safeguarding personal data ownership, privacy, and usage and attributing value to the ones who produce it. Interesting approaches of personal data management therefore make use of smart contracts and distributed ledger technology.

For the purposes of a personal data sharing platform, it should be investigated if and how to ensure the electronic identification and to get the verifiable credential (on the basis of a national digital identity), where necessary for accessing to online public services.

The eIDAS[33] Regulation states that the processing of personal data must be carried out in accordance with the GDPR and respecting its principle of

[32] UNESCO, Recommendation on the Ethics of Artificial Intelligence, 2021. Available at: https://en.unesco.org/artificial-intelligence/ethics (accessed Jul. 29,2022)

[33] Alamillo Domingo, I., "SSI EIDAS Legal Report - How EIDAS Can Legally Support Digital Identity and Trustworthy DLT-Based Transactions in the Digital Single Market", 2020.

confidentiality and security of processing: as clarified in its Recital 11, the authentication for an online service should concern processing of only those identification data that are adequate, relevant, and not excessive to grant access to that service online.

In case the platform concerned foresees to use electronic identification for its users, either natural or legal persons, this eIDAS Regulation can become applicable for its services and should be investigated especially in the context of the wallets and the smart contracts. Its electronic identification (eID) tools can be used for the identification of users, as they broadly offer enhanced security and accuracy, swifter, and less costly processes, while they may mitigate risk of fraud, identification theft, and legal challenges.

On the other hand, the concept of self-sovereign identity (SSI)[34] could also present advantages for the purpose of a personal data platform deployment and use and should therefore be investigated, including its compliance with eIDAS.

Sovrin[35] argued that the "self-sovereign identity (SSI) is a term used to describe the digital movement that recognises that an individual should own and control their identity without the intervention of administrative authorities. SSI allows people to interact in the digital world with the same freedom and capacity for trust as they do in the offline world". Furthermore, "Blockchain and SSI are natural complements, making the perfect symbiosis": the user is able to individually create and manage his/her identify thanks to the use of distributed ledger technologies (e.g., blockchain), without the involvement of a third party, but often making use of the "decentralised identifier" (DID) associated with an entity. Such entity using SSI to authenticate itself can be an individual (natural person), and, therefore, in this case, the DID usually relates to an identified or identifiable person (thus being personal data).

The SSI enables sovereignty for individuals over their digital assets and credentials, often by using digital wallets. In case the individual presents such assets and credentials to a third party to prove ownership, the public, decentralised, and immutable registry (such as a blockchain network) can be employed: the cryptographic proofs of the asset or credential were registered and are kept in a standardised and trustable way.

Available at: https://joinup.ec.europa.eu/sites/default/files/document/2020-04/SSI_eIDAS_legal_report_final_0.pdf (accessed Jul. 29, 2022)

[34] Allende Lopez, M., Self-Sovereign Identity: The Future of Identity: Self-Sovereignty, Digital Wallets, and Blockchain, 2020, http://dx.doi.org/10.18235/0002635

[35] Sovrin, Trust Assurance Framework, 2019. Available at: https://sovrin.org/wp-content/uploads/Sovrin-Trust-Assurance-Framework-V1.pdf (accessed Jul. 29, 2022)

Nonetheless, the question whether eIDAS is already suitable for SSI and blockchain technology is still open, as well as whether, on the one hand, the smart contracts could be considered electronic documents and, on the other hand, the means used to sign blockchain transactions could be considered electronic signatures, with all the legal consequences it implies. Some scholars[34] argue that the eIDAS Regulation will need some adjustments to become the legal and trust framework for SSI in the European Union: it was created as a legal framework supporting a digital identity metasystem mainly based on delegated authentication, which is more limited than the self-sovereign approach which enables, among other things, pseudonymity and selective disclosure mechanisms.

In the US system, the situation is not exactly the same and some authors underlined that blockchain transactions can constitute, or evidence, electronic signatures and that, virtually, all transactions stored on a blockchain, and retrievable in perceivable form, constitute an electronic record under the US law.[36,37]

In conclusion, it is not fully clear whether for the purposes of a personal data sharing platform, it should be ensured (and how) the electronic identification and it should be necessary to get the verifiable credential (on the basis of a national digital identity), where necessary for accessing to online public services.

On the other hand, from the viewpoint of the smart contract itself, often used in the personal data platform to give the compensation for the sharing of own personal data, the debate is still ongoing whether and to what extent and conditions, these can give rise to legally binding and enforceable contracts and whether this necessarily requires the identification of the individual pursuant to eIDAS.

The smart contract satisfies the elements of a contract under several national laws, such as Spanish Civil Code, and, therefore, smart contract code represents a valid mechanism to define the parties' contractual rights and obligations as a matter of contract law in many jurisdictions. Therefore, "under certain circumstances, and if so decided by the parties, smart contracts can fulfill the elements of a legally binding contract under common law and civil law systems".[38] Though the parties may act pseudonymously,

---

[36] U.S. Government, Public Law 106 - 229 - Electronic Signatures in Global and National Commerce Act (ESIGN), 2000. Available at: https://www.govinfo.gov/app/details/PLAW-106publ229

[37] Therefore, under certain legislation, blockchain platforms may constitute or store electronic records and electronic signatures and thus may be used to evidence, or give effect to, electronic or smart legal contracts

[38] Smart Contract Alliance, "Smart Contracts: is the Law Ready?" 2018. Available at: https://digitalchamber.org/smart-contracts-whitepaper/ (accessed Jul. 29, 2022)

it is necessary a link (including off-chain) to their real identity to provide for valid consent, which is a crucial element of a contract under several national systems. However, even if its deployment does not give rise to a legally binding contract, the smart contract may still affect legal relations (either between the parties or with third parties) and therefore may have legal effects.

At the same time, both smart contracts and conventional natural language contracts can coexist in relation to the same (or related) subject matter and create together the entire legal framework within which a smart contract operates. This is the case of the so-called "external smart contract", where "the code does not form the entirety of the parties' legal agreement, but merely automates the performance of some of its terms".[37] The code merely automates the performance of some of the conventional contract's terms. In this case, the legal relationship is intended to be governed by the natural language version of the contract, rather than by the code. In the internal model, on the contrary, the code could either encompass the entire agreement between the parties, or, alternatively, could form only an integral part of the legally binding contract (rather than the entirety of the contract), and would supersede any other clauses written in natural language: the code would be given legal effect and is an integral part of the agreement.

Principally, it is necessary to refer to the governing law applicable to the smart contracts in order to determine whether these give rise to legally binding contracts, whether personal identification is necessary or not according to eIDAS, as well as to evaluate the effects of the DTL/blockchain, and, ultimately, to ensure that the model chosen meet local law requirements. However, considering that the DataVaults offering can constitute an electronic registered delivery service according to eIDAS (Article 3, (36) eIDAS), such regulations and the obligations established for the providers of such services have to be taken into account in the design, development, and future use of personal data platform.

15.10 DataVaults as a Flagship Initiative for Personal Data Sharing Under User Control and Benefitting All the Actors Involved: Experiences and Lessons Learnt

The DataVaults project is directed to rejuvenate the personal data value chain by delivering a framework and a platform having personal data, coming from diverse sources (wearables, web APIs, smart home sensors, personal data records, etc.) in its centre. Secure, trusted, and privacy-preserving mechanisms have been designed to allow the individuals to take ownership and control of their data and share them at will, through flexible data sharing solutions and fair compensation schemes with other entities (companies, public bodies, or other organisations).

DataVaults aspires to become one of the flagship personal data platforms in the European landscape, characterised by full respect of GDPR provision and satisfaction of the privacy and trust consideration of users, with a novel, fair, and understandable value compensation mechanism to data owners.

Therefore, the whole Consortium paid great attention to tackle any potential ethics issues raised by the platform's validation and future operation in order to give rise to a technology respectful of the data subjects' privacy and dignity and capable of prioritising human wellbeing and flourishing.

For this purpose, they elaborated the DataVaults Ethical Policy at the beginning of the project and adhered to it, conducted an in-depth regulatory review, elicited a set of legal and ethical requirements and related guidelines and recommendations for the overall DataVaults cloud-based platform and its components, as well as the Personal App and the demonstration activities. They also followed (and are going to conduct again at the end of the project) an Ethics and Data Protection Impact Assessment methodology, besides capturing the citizens' perspective through dedicated interactive channels.

The following outlines the activities performed and outcomes achieved by the Consortium in order to adhere to the highest ethical standards and comply with the legislation, *in primis* the Data Protection Law (especially the GDPR).

15.11 Case Study: Approach and Legal and Ethical Requirements for DataVaults Ethical Policy

The DataVaults Ethical Policy has been conceived and implemented to ensure the legitimacy and fairness of project technologies and demonstrations. It depicts the ethical procedures and responsibilities, including those relevant for human participation and personal data collection and processing in the demonstrators, besides identifying the oversight responsibilities (with the appointment and involvement in project's activities of the DataVaults Ethics & Data Protection Officer and DataVaults Ethical Board) and setting the basis for the comprehensive Data Protection Impact Assessment methodology used during the demonstrators operations. The Policy also drew the roadmap for the implementation of ethics-related activities within the project.

The Policy is driven by the Fairness & Privacy-by-Design-and-by-Default enriched with the Protection Goals Approach, adopted for analysing the composite regulatory landscape, for deriving the legal and ethical requirements, as well as for providing recommendations and insights on how to face the identified boundaries and constrains. This approach is functional to ensure that the research activities, results, and validation activities are legally compliant and ethically sound. First of all, GDPR itself sets forth among the principles relating to processing of personal data the so-called "Fairness Principle". Fairness, which can be explained through the concepts of loyalty and good faith to be respected in all the steps of any personal data processing, requires that personal data must be used in a fair way, avoiding to process in a way that is unduly detrimental, unexpected, or misleading to the individuals concerned or that could have adverse impact on them. The "Fairness by Design" has identified a straightforward requirement for DataVaults technology in order to ensure that individuals' privacy and real control over their data. The procedural dimension of the fairness entails the effective exercise of the data subjects' rights (rectification, erasure, object, etc.), whilst its substantive dimension implies moving towards the equal and just distribution of benefits and costs, without unfair bias, discrimination, and stigmatisation for individuals and groups. This is linked with another high-level ethical requirement, the "sharing the wealth" paradigm,[39] aligned with the vision of a win–win data sharing ecosystem fostered by the Big Data Value Association[40] as a contribution to help unlock the social value of personal data, going beyond user consent for fostering individual human empowerment and flourishing, as well as the common good of society and businesses' interests. The DataVaults Consortium followed this approach and directed its efforts to promote the alignment of its research and outcomes with social needs and expectations, also in view to strengthen the societal uptake of DataVaults cloud-based platform, given that high ethical standards generally imply public trust.

This approach might be particularly relevant also in an urban and public environment since it supports the identification, on a case-by-case basis, of the proper balance between competing interests and encompasses societal fairness, based on equal opportunities and on the need to avoid that individuals are deceived or unjustifiably impaired in their freedom of choice. In view of fully ensuring the fairness of the technological artefact, it is advisable

---

[39] Bormida, M.D., "The Big Data World: Benefits, Threats and Ethical Challenges", Iphofen, R. and O'Mathúna, D. (Ed.) Ethical Issues in Covert, Security and Surveillance Research (Advances in Research Ethics and Integrity, Vol. 8), Emerald Publishing Limited, Bingley, pp. 71–91, 2021. https://doi.org/10.1108/S2398-601820210000008007

[40] BDVA Position Paper "Towards a European Data Sharing Space - Enabling data exchange and unlocking AI potential", 2019. Available at: https://www.bdva.eu/sites/default/files/BDVA%20DataSharingSpace%20PositionPaper_April2019_V1.pdf (accessed Jul. 29, 2022)

to investigate several dimensions and take into account different perspective, for instance, focusing the attention on different kinds of compensation mechanisms, besides data monetisation schemes, such as other rewarding incentives, so that the different brackets of the population will be encouraged to share data.

The chosen approach strongly relies also on human-centricity. Exploring and deepening individuals' viewpoint was considered essential by the Consortium for effectively adhering to the chosen Ethics, Fairness & Privacy-and-Security-by-Design-and-by-Default Approach and for contributing to build a win–win data sharing ecosystem.

For this reason, in order to capture citizens' perspective, expectations, needs, and concerns on personal data sharing, the Consortium conducted a survey directed to individuals in their role of data owner. Results from the survey provided an understanding of:

- attitudes towards personal data sharing;
- data retrieval, storage, and deletion;
- privacy preservation on the shared data;
- compensation mechanisms;
- control and informed consent.

These results, as well as of the other stakeholder engagement activities, were key for driving the design, development, and deployment of the Personal Data Platform and App planned in DataVaults, whilst also providing important indications for the future progress of the Personal Data Market in Europe.

This attention to the individual is also consistent with the EC strategy and vision[41] directed to put people first in developing technology and to promote European values and rights in any design, development, and deployment of the technology in the real economy.

The Ethics Board offered guidance, advice, monitoring, and recommendations for future work, mainly with respect to ethics and privacy, whilst the Ethics and Data Protection Officer (EDPO) mainly supported the partners in ethics compliance and in the handling and management of personal data in accordance with the existing provisions of GDPR and other relevant EU and national legislations, providing guidance and advice, training of researchers, assisting in ethics risk assessment, and supporting in relation to the Ethics and Data Protection Impact Assessments.

---

[41] See, for instance, COM/2020/66 final, "A European strategy for data" (ref. number 4).

On the other hand, the Policy also drew the ethical procedures for the human involvement and personal data collection and handling in the demonstration activities, since individuals will be involved in the pilots and their personal data, coming from diverse sources (sensors, IoT, wearables, data APIs, historical data, social network data, activity trackers, health records, demographic profiles, etc.) were gathered, processed, and shared. These procedures include those used to identify/recruit research participants, as well as the high-level description of the informed consent procedures for the participation of humans and personal data collection and processing, including also the sample of the informed consent/assent forms and information sheets distributed to the research participants. Such samples were fine-tuned and adapted by each relevant demonstrator, taking into account the specific context, technologies, and scenarios, with advice available from the EDPO where required.

The Ethical Policy guided the ethics-related work carried out by the project partners, both in the technical work-packages where the technological assets are designed and developed, and in the demonstration activity, where the results are assessed.

In particular, the Policy is strongly interrelated with the legal and ethical requirements elicitation. At an early stage of the project, the legal and ethical requirements for the design, development, and validation of DataVaults cloud-based platform and Personal App were set, alongside the future operation of them, clearly laying out a first guideline for legal compliance and ethically sound activities and results, without forgetting checkpoints. The initial requirements list was extended taking into account the enriched legal review, where additional areas of law were analysed, as well as the regulatory reforms under development and their accompanying documents.

All these requirements were elicited adopting a systematic and holistic approach, driven by Fairness & Privacy-by-Design-and-by-Default enriched with the Protection Goals method and relying on the analysis of the regulatory landscape and the factual analysis of the privacy-relevant properties and personal data collection, processing, and sharing in each service and tool, including details on the data categories, data sources, and purposes of processing.

Some of the requirements are binding (when directly deriving from the legislation, such as GDPR), whilst others, where not directly imposed by the legislation, have to be interpreted more than recommendations or preferable requirements. Some requirements, being quite challenging, need to be assessed with a certain degree of flexibility, taking into account the state-of-the-art of the technological developments and the risk-based approach fostered by GDPR itself: in other words, this demands for a certain degree of flexibility in the assessment of the adequateness of measures and technological solutions, to be specifically established on a case-by-case basis, considering a set of circumstances rotating around the severity of the risks and the reasonable efforts to face with them.

The nature of the requirement is clearly stated in the description of each of them and they are provided in a table format, in order to facilitate the quick understanding and reference to them by the technical team. Furthermore, in order to promote the operationalisation of the requirements, additional notes, recommendations, and guidelines were provided.

The fulfilment of the requirements regarding the DataVaults technology ensures that it is legally compliant, ethically sound, and gives rise to a trusted, secure privacy-friendly enhanced (holistic) data sharing solution. The assessment of the compliance with these requirements was conducted in a triple iteration, respectively, concerning the alpha, beta, and final version of the DataVaults technology (in particular, the platform).

15.11.1 Ethics and data protection impact assessment methodology

An important element of DataVaults Ethical Policy was the definition and implementation of the Ethics and Data Protection Impact Assessment Methodology for the demonstrator cases, functional to the assessment of risks for individuals' rights, freedoms, and wellbeing, for ensuring compliance with the data protection law (GDPR and national regimes), and ethical mandates.

This methodology regarding the risks for the personal data was conducted following the indications of Article 35 section 1 GDPR, taking into account the nature, scope, context, and purposes of the processing operations in each demonstrator in view of evaluating their impact on the protection of personal data, to identify and reduce the data protection risk[42] and the likelihood of privacy harms to individuals, as well as to identify and put in place the appropriate technical and organisational measures to tackle with/mitigate such risks.

A model inspired by the ISACA Model[43] was adopted for conducting such data protection assessment, which maps the 14 ISACA privacy principles

---

[42] The concept of risk is clarified in Recitals 75–79 of the GDPR.

[43] ISACA, "GDPR Data Protection Impact Assessment", 2017. Available at: https://isaca-gwdc.org/wp-content/uploads/2018/01/GDPR_res_eng_0917.pdf (accessed Jul. 29, 2022)

to the specific GDPR requirements and therefore allows an easy integration with any additional privacy impact assessment (PIA) standards required for other possible multiple privacy principles relevant for the demonstrators. Furthermore, this model is well aligned with the protection model focused on individual privacy and user control and efficaciously supports accountability, representing a useful instrument for the demonstrators to showing commitment and due diligence in taking adequate actions to ensure full compliance on an ongoing basis.

The demonstration sites elaborated their own EDPIA in conjunction with the respective technological supporting partners and the overall technical team of the project. It considered the specific technologies (like services and components) relevant to their context, the data lifecycle and each use cases scenarios, as well as their own privacy and security policies/practices.

Furthermore, in order to adequately cover also the ethical dimensions and to assess to what extent the principle of fairness has been operationalised in each of the demonstrator, the model inspired by the ISACA scheme was enriched with the Data Ethics Canvas.[44] This tool was elaborated by the ADAPT Centre for Digital Content Technology on the basis on the original Business Model Canvas by Alex Osterwalder.

This model consists in a useful tool giving a higher level framework to develop ethical guidance that suits any context and to assess the ethical implications of any project, thereby allowing to be more trustworthy with data processing.

The Data Ethics Canvas is capable of helping those who collect, share, and use data in identifying and managing ethical uses, both at the start of the initiative which imply data collection/processing and throughout[45] the implementation of the initiative. On the other hand, thanks to it, the data seekers are supported in putting in place practices ensuring that the way the data is collected and used is trustworthy and ethical, beyond legal compliance.

The Open Data Institute's Theory of Change is strongly consistent with the DataVaults' vision and with the Citizen Control of Personal Data Initiative

---

[44] Reijers, W., Koidl, K., Lewis, D., Pandit, H.J., Gordijn, B., Discussing Ethical Impacts in Research and Innovation: The Ethics Canvas. In: Kreps, D., Ess, C., Leenen, L., Kimppa, K. (eds) This Changes Everything – ICT and Climate Change: What Can We Do?. HCC13 2018. IFIP Advances in Information and Communication Technology, vol 537, 2018. Springer, Cham. https://doi.org/10.1007/978-3-319-99605-9_23

[45] Open Data Initiative (The ODI), "Helping organizations navigate ethical concerns in their data practices", 2017. Available at: https://theodi.org/article/the-data-ethics-canvas-2021/ (accessed Jul. 29, 2022)

within the Smart Cities Marketplace – Citizen Focus Action Cluster: "We want people who steward data, and people who create things with it, to act in ways that bring about positive impacts. Ethical use of data helps to improve trust and bring about the best economic and social outcomes. We want to avoid a future where data is feared or hoarded. We want data to work for everyone".[46]

The Ethics and Data Protection Impact Assessment was conducted in each project's pilot through a questionnaire comprising elements coming both from the ISACA Model and from the Data Ethics Canvas. Strong reference was made, besides internal own policies, to the legal and ethical requirements set by the project itself. The EDPIA represented a key tool for ethical assessment and compliance in DataVaults and can be easily replicable, with the necessary adaptations, for use in other contexts like the public sphere.

15.12 Conclusion

In view of strengthening the development and growth of the data economy also in relation to personal data, it is key to foster the adoption of trusted and secure personal data platforms capable of handling back control over the use of personal data to individuals giving them actual benefits, not-necessarily financial. Future efforts should be directed towards building a win–win data sharing ecosystem in order to unlock the social value of personal data, going beyond user consent for fostering individual human empowerment and flourishing, as well as the common good of society and businesses' interests. In alignment with the EC's vision of personal data sharing that includes benefits for all the actors in the value chain, trusted, secure, and value generating data management and sharing platforms for personal data should be encouraged to the extent that they allow stakeholders' collaboration for supporting their own goals and operations, as well as allowing further stakeholders, such as the local communities and local authorities, to offer new socially and environmentally sustainable solutions and business models.

In this environment, on the other hand, the technologies should move to regain the trust of individuals when it comes to data sharing, leaving the control in their hands for deciding how, how much, and in which manner they would like to share their data, whilst at the same time guaranteeing their privacy and with adequate security levels, as well as ensuring

[46] Open Data Initiative (The ODI)– Our theory of change. Available at: https://theodi.org/about-the-odi/our-vision-and-manifesto/our-theory-of-change/ (accessed Jul. 29, 2022)

fair share of the value that their data generates, also in case of secondary operations.

In other words, human-centricity should be at the centre of the future technological developments and their operation when it comes to data sharing. Prioritising human wellbeing and fundamental rights and putting people first in the data-driven economy are expected to contribute to rebuild public trust and, therefore, societal acceptance of such innovations. This is also aligned with the Communication "2030 Digital Compass: The European Way for the Digital Decade".[47] Its Vision for 2030 relies on empowered citizens and businesses: "the European way to a digitalised economy and society is about solidarity, prosperity, and sustainability, anchored in the empowerment of its citizens and businesses, ensuring the security and resilience of its digital ecosystem and supply chains" with four cardinal points for mapping the EU's trajectory.

Personal data sharing platforms, like DataVaults cloud-based platform, capable of fully embracing this vision and the promotion of the EU's fundamental values (including protection of privacy) are expected to contribute to the creation of a single market for data that will ensure Europe's global competitiveness and data sovereignty. This will allow an increased amount of data made available for use in the economy and society, but at the same time safeguard individuals by effectively empowering them to exercise their rights with regard to the use of the data they generate and to decide at a granular level about what is done with their data, moving towards "personal data spaces".

[47] COM/2021/118 final, Communication from the Commission to the European Parliament, the Council, the European Economic and Social Committee and the Committee of the Regions, "2030 Digital Compass: the European way for the Digital Decade". Available at: https://eur-lex.europa.eu/legal-content/en/TXT/?uri=CELEX%3A52021DC0118

16

Data-Driven and Citizens' Inclusive Smart Cities: Top-Down and Bottom-Up Approaches to Tackle Societal and Climate Challenges

Maria Sangiuliano

Research Director, Smart Venice, Venice, Italy
Email: maria.sangiuliano@smartvenice.org

Abstract

This chapter initially covers the main work which is the Citizen Focus Action Cluster in the Smart Cities Marketplace. It draws attention to the resources that are available for cities to use in relation to collaborating with other cities when addressing inclusion issues in designing their urban strategies.

It addresses the crucial role of data in citizen engagement in the development of smart cities and does this by presenting three case studies.

It concludes with a reflection on the future trends which are worth following covering policy and technology developments in this field.

16.1 Introduction

A citizen-centred approach has become an inherent part of the definition of intelligent and smart cities in the past few years, although it is taken into effective consideration at different degrees, from mere statements and communication-oriented approaches (buy-in of citizens as technologies' users/adopters) to more or less "critical" and "transformative" visions that foresee to engage with citizens playing active roles in all phases of a smart cities project, encompassing planning/designing implementation and evaluation phases, thus contributing to the renewal of urban democratic processes.

DOI: 10.1201/9781003399384-17

The entire field of promoting smart cities has been indeed subject to criticism of it being far too "submissive" to market forces and the deployment of technologies at the urban scale by replicating discourses on "creative cities" that would end up (mostly) benefitting citizens of high–middle classes and reproducing existing inequalities and divides. In the work of the EIP smart cities and communities (re-named Market Place[1] after 2020), we have aimed at facilitating dialogue and discussions among the broadest possible spectrum of stakeholders interested into overcoming "tech-centred" approaches so as to counteract the drift of the smart cities discourse, towards merely market-driven directions: for this purpose, a dedicated "citizen focus" action cluster[2] has been set up and developed through the years, focusing on citizen engagement in smart cities, which we humbly believe have contributed to make the discursive and empirical frameworks of smart cities all in all more inclusive.

This short chapter has a three-fold aim: it intends to present the main frameworks and action lines of the Citizen Focus Action Cluster since its creation to date, and to point at the key resources that we believe can still be a point of reference for cities that want to work with other cities when thinking of their own urban innovation processes in an inclusive manner. On the other hand, and coherently within the context this volume provides, we wish to emphasise the crucial role of data in citizen engagement within smart cities projects and to present three concrete case studies/examples. Finally, we reflect on future trends to follow up and keep track of in the next few years in view of the most recent policy and technology developments in the field.

16.2 Sharing and Networking on Citizen Engagement in Europe. Resources and Lessons Learnt from the Citizen Focus Action Cluster of the Smart Cities Marketplace

The Action Cluster had started in 2014 by facilitating networking and mutual learning among 61 commitments submitted by about 100 organisations and projects, several of those being funded by the European Commission. After having distilled the good practices and the principles emerging from the bottom-up process harvesting knowledge from cities, researchers, and

[1] https://smart-cities-marketplace.ec.europa.eu/

[2] https://smart-cities-marketplace.ec.europa.eu/action-clusters-and-initiatives/action-clusters/citizen-focus

companies into a first booklet, the Cluster has converged on one main communication campaign with an Inclusive Smart Cities Manifesto[3] that has positioned citizen engagement at the core of the smart cities discourse, linking it closely to the design phases, presenting co-design and co-creation as some of the methods to pursue this, so as to overcome the narrower and top-down interpretations of citizen engagement as ways to convince/educate citizens that have featured in its initial steps (as highlighted by Cardulllo and Kitchin, 2019[4]). Moreover, the Manifesto has represented a first move towards overcoming a "universalistic" and generalising definition of "citizen" that is too often concealing an idea of a citizen as white-young or middle aged, middle or upper-class, CIS-gender or male, and able citizen.

The Manifesto has put forward a vision of inclusiveness within its first statement "in the design of smart cities solutions, it is crucial to use the appropriate means to engage and empower population groups difficult to reach such as people experiencing poverty and/or social exclusion, younger and older people, migrants, people with disabilities, and aim at gender balance in participation and engagement". Finally, the Manifesto has conveyed the message that citizen engagement is boosted by data and can foster data generation by citizens, as it encouraged cities to "Promote the use of open data and/or an appropriate access to data by citizens, developers, start-ups and engaging citizens in the evaluation of urban policies by applying Open Government practices integrating feedback loops to renew and improve services, ensure privacy by design". Presented in a high-level launch event in Brussels, translated in many European languages, and signed by more than 200 organisations across Europe, the Manifesto has supported a paradigm shift that was becoming mature and that has been continued in the forthcoming activities of the Action Cluster Citizen Focus.

Also, to confirm our non-reductionist vision and diversified vision of citizen engagement, we have put efforts in promoting participatory budgeting (PB) as a methodology that enables citizens to decide on how to allocate portions (although small ones) of the city's budget: in fact, PB integrates participatory and consultative elements, crowdsourcing of project ideas, with deliberation and voting on a selection of the crowdsourced project ideas. We have curated and disseminated knowledge and good practices on participatory

---

[3] https://smart-cities-marketplace.ec.europa.eu/sites/default/files/EIP-SCC%20Manifesto%20on%20Citizen%20Engagement%20%26%20Inclusive%20Smart%20Cities.pdf

[4] Cardullo, P. and R. Kitchin (2019). Smart urbanism and smart citizenship: The neoliberal logic of 'citizen-focused' smart cities in Europe August 2019. Environment and Planning C Politics and Space 37(5): 813–830.

budgeting as well as the online platforms that enable them, acknowledging how back in 2016, already over 8 million European citizens had been actively involved in participatory budgeting initiatives (EPRS, 2016, p. 8[5]) and its relevance and strong potential in the context of smart cities planning and implementation. Many cities have been increasingly using online digital tools to implement participatory budgeting, either through in-house designed or open source or proprietary platforms. Furthermore, it is proven how digitalised versions of PB can both empower citizens and acknowledge their role in decision-making processes and achieve "smart literacy" goals.

Participatory budgeting can be experimented and piloted with thematic focus on areas of intervention which are particularly relevant for smart cities, such as smart mobility, energy efficiency, waste management, etc. (i.e., Capaccioli *et al.*, 2017[6]), and this type of experiences has been reported and analysed. The work carried out on PB in the Cluster fed into the chapter on citizen engagement that was included in the Smart City Guidance Package.[7] Whilst promoting the good practices and the online platforms in use from several cities (Barcelona, Madrid, and Paris, among others), we have acknowledged as the most promising ones those experiences such as in Barcelona, where participatory budgeting has been part of a broader set of policies aimed at re-interpreting a smart city definition around notions of technological sovereignty and politics around the "right to the city". In Barcelona, the Decidim platform[8] has been the core of the participatory budgeting experiment, "a digital infrastructure for participatory democracy", a "public-commons" project mostly financed and made possible by the city but designed and maintained by an open-source community. It is used to consult and co-create with citizens on the specific topics/policies/regulations at stake at a given time, and the digital component of collecting ideas and feedback on the platform is always accompanied, pandemic conditions permitting, by in-presence meetings.

Critical aspects that the focus groups and webinars on PB which we have organised, have shown that there is a geographical gap as this practice

---

[5] European Parliament Research Services (2016). Participatory Budgeting. An Innovative Approach https://www.europarl.europa.eu/RegData/etudes/BRIE/2016/573894/EPRS_BRI(2016)573894_EN.pdf

[6] Capaccioli, A., Poderi, G., Bettega, M. and V. D'Andrea (2017). Exploring participatory energy budgeting as a policy instrument to foster energy justice. Energy Policy, vol. 10, pp.621–30.

[7] https://www.researchgate.net/publication/343615678_Smart_City_Guidance_Package

[8] https://decidim.org/

and its variants and platforms are much less present in eastern European countries. Other hindrances that have been shared within the Citizen Focus Action Cluster community, included some of those also recalled by a recent study (Radma-Liiv, 2022[9]) stemming from the Horizon 2020 Tropico project. This compared e-participation experiences in six European countries. A careful design and implementation is required as PB experiments require the collaboration of multiple internal and external stakeholders; so any flaw in the process can lead to delayed decisions, increased conflict, and distrust from participants. The research, mutual learning, and networking activities of the Citizen Focus Action Cluster have along the years encompassed many more topics and areas than the ones mentioned above, through the establishment of so-called "Initiatives" (or cluster sub-groups): citizen-centric approach to data, Citizens' Control of Personal Data,[10] Policy Labs,[11] and the creation of the Citizen City[12] toolkit are among the most prominent ones. Most of the knowledge that has been generated has converged into the "Citizen Engagement Solution Booklet"[13] that stemmed in 2021 from the collaboration between the Smart Cities Information System and the Action Cluster: in fact, in the booklet, we defined citizen engagement as "a way of actively involving citizens in the city's decision-making processes, that can help to address these needs. Citizen engagement exists in many different forms, ranging from merely influencing and informing people, to real participation and actual decision making. The set-up of such an engagement process could be initiated by the city or its citizens, and in its most thorough form is organised by collective effort".

The booklet provides a substantial and informative set of resources accompanied by concrete cases and good practices from cities covering the different steps of setting the goals for citizen engagement and co-creation of urban innovations, designing and implementing the interventions, and scaling results up to the policy levels. It also encompasses a variety of sectors, from mobility to energy communities and waste management and more.

[9] Randma-Liiv, T. (2021). Organizing e-participation: Challenges stemming from the multiplicity of actors. Public Administration, 1–17.https://doi.org/10.1111/padm.12788RANDMA-LIIV17

[10] https://smart-cities-marketplace.ec.europa.cu/action-clusters-and-initiatives/action-clusters/citizen-focus/citizens-control-personal-data

[11] https://smart-cities-marketplace.ec.europa.eu/news-and-events/news/2018/policy-labs-building-future-policy-making

[12] https://smart-cities-marketplace.ec.europa.eu/action-clusters-and-initiatives/action-clusters/citizen-focus/citizencity

[13] https://smart-cities-marketplace.ec.europa.eu/insights/solutions/solution-booklet-citizen-engagement

A main obstacle that we have detected on the way towards "mainstreaming" a citizen engagement approach to smart cities initiatives refers to the lack of resources, meant as both financial and human: those cities that are or have been partners within Horizon 2020/Europe funded projects have more frequently incorporated citizen engagement practices (although with different "intensity"); when external funding is not available, the needed investment might be lacking. Funded projects often function as learning pathways where allocated resources allow for input from specialists/experts partners to guide the cities' efforts from a methodological point of view, while at the same time building internal skills and capacities; yet, restructuring the processes of a public administration and shrinking budgets are often mentioned as impediments to allocating staff and resources dedicated to engagement and participatory processes that feed into urban innovation policies and projects.

16.3 Good Practices. Citizen Generated Data to Improve Urban Innovation and Smart Cities Policies. Top-Down and Bottom-Up Approaches

The good practices we present and describe in this section have been selected with the rationale of, on the one hand, focusing on a more top-down approach to the engagement work that can be done relying on data. This is the case for Camden, London (see Section 3.3) and for Sofia (see Section 3.11) through its Policy Cloud H2020 project[14] when decision is made by the city administration to put in place a project and/or make use of a technology that makes the most out of data for the sake of urban innovation policies and their evidence-based qualities, while preserving citizen's right to privacy. A complementary comprehensive review of citizens' generated data practices as forms of "data donation" that can benefit public policies is to be found in a JRC report (Ponti & Craglia, 2020[15]): it documents 18 European projects involving citizen-generated data via digital technologies (low-cost sensors and accessible digital technologies). These have proven to re-configure the relationship between citizens and the public sector, giving the former ones increased agency and control so that they undertake a "change-agents" role: this is particularly important in times of serious challenges related to the

[14] https://policycloud.eu/

[15] Ponti, M. and Craglia M. (2020). Citizen-generated data for public policy. A brief review of European citizen generated data projects, European Commission, Ispra, 2020 JRC120231.

climate crisis that make a change agenda urgent and enhanced accountability and transparency from local authorities a must.

On the opposite, the other set of good practices featured in this section was chosen as they are more bottom-up in terms of their origin, management, and development, and refer to the use of crowdsourced data and information from citizens to feed into web and mobile applications that are aimed at addressing societal challenges. The Europe- and India-based projects that we will refer to show interesting and challenging aspects as far as harnessing citizen-generated data to contribute to solving pressing societal matters, such as gender-based violence and increasing the wellbeing and perception of safety in public spaces.

16.3.1 Harnessing open data for evidence-based urban policies – the Camden and Sofia use-cases

The Horizon 2020 Policy Cloud project on cloud-based solutions for data-driven policy management has a whole field of experimentation and piloting on policies for citizens. This has been utilised by the Camden Borough in London, UK and by Sofia, the capital city of Bulgaria, which is a pilot site for the project. To build internal legitimacy for the processes described below, Policy Cloud has leveraged on recent EU policies such as the Communication Towards a Common European Data Space (2018)[16] on fostering access and re-use of public and publicly funded data as a cornerstone of an European Data Space as well as the former Revised PSI (Public Sector Information) Directive from 2013.[17]

The Camden Borough in London was one of the Policy Cloud piloting partners, and it applied the use of the project platform and data analytics technology for predicting risks related to unemployment in its 210,100 population. Starting from an existing commitment to publish open data in adherence to the International Open Data Charter, the Council is relying on hundreds of published datasets from offices that manage unemployment schemes and allowances but also housing and social welfare more broadly, to highlight trends. Such a work is based on data and is underpinned by the notion that residents have the right to access data which does not compromise individual privacy. The ultimate objective is to identify impacting factors and thus informing evidence-based policy choices that have increased feasibility and achieve greater legitimacy. The Council checks on KPIs and statistics

[16] https://digital-strategy.ec.europa.eu/en/news/communication-towards-common-european-data-space

[17] https://digital-strategy.ec.europa.eu/en/policies/psi-open-data

based on the number of citizens claiming governmental aid when seeking work: through data analytics performed by the platform developed by Policy Cloud. Age groups most affected by unemployment are identified and issues with discrimination related to gender bias or other biases on the ground of ethnic origins, as well as the specific areas of the borough that might be more heavily affected enable policy choices to become more responsive to the needs of specific populations. The pilot has shown several advantages of the platform and proven its accessibility, due to the lack of need for infrastructure to run the tool and its adaptable features.

Sofia has instead used the concept of improving urban policy-making through the analysis of crowdsourced data by using data from direct communication with and from citizens and the city administration itself. The capital of Bulgaria with its population of about 1.8 million inhabitants has sought for improving the overall urban environment of the city, via assessment and validation of urban policies through big data analytics. Analysis of big data has been sourced from different sectors, such as: transport, parking and road infrastructure, waste collection, cleanliness of public spaces, air pollution, and violation of public order. The main existing data source for the cloud-based Policy Cloud platform has been the citizens' contact centre ("SofiaCall" service), which has been operational since 2014 and facilitates direct communication from citizens, industry, and institutions wanting to signal non-urgent deviations in the urban environment. Citizens use the contact centre to post online signals (as are quite typical for this type of services, free text and pictures can be uploaded, categories/labels assigned and each signal is geo-located). Call Sofia contains 300.000 signals in 90 categories, 70.000 in 2019, each one manually processed by officers.

Analysing the territorial distribution of signals by category through the Policy Cloud platform enables municipal and district administrations to identify problems, issues, and behavioural trends in the urban environment. Monitoring and control of the services are also made possible: risks are identified and preventive actions can be envisaged, as well as adjustments being made to policies. An additional crowdsourced database that has been used in the project is the "Air Things Platform"[18] for air-quality monitoring (while all scenarios use data from Sofia Call, one uses also Data from Air Things). What happens with Policy Cloud making use of crowdsourced data is that it produces meaningful and structured results starting from unstructured data: incidents and problems are categorised and analysed by type, area, frequency,

[18] http://www.interreg-balkanmed.eu/article/137_AIRTHINGS-Project---Open-Data-Platform-

and year/month, and patterns are identified that enable predictions and forecasts. Re-usable analytics tools are made available to the city, as well as a "Data MarketPlace" that is respectful of privacy and GDPR regulations.

Visualisation tools embedded in the platform proved to be crucial for buy-in of decision-makers and policy officials (bar charts, heat-maps, spider charts, etc.). Interestingly, a co-creation process was organised from the design and validation phases of the scenarios for use of the platform itself with feedback collection from different stakeholders inside and outside the city, involving the innovation hubs in the city, such the Gate Institute Centre of Excellence on Data and Artificial Intelligence within the Sofia University.

16.3.2 Crowdsourced data for enhancing safety perception in public space and transport

Safecity was established in December 2012 as an immediate response to a gang-raping episode ending up in femicide of the raped and brutalised victim who got national and international media attention in Nirbhaiya, India. The debate focused on the opportunity for women to avoid going out in the evening as a possible solution; so the promoting NGO took on an empowerment-oriented approach instead, creating means of reporting the places where women face harassment or violence so that "hot spots" of incidents can be viewed on a map. The data collected on the "Safecity.in[19]"platform consist of individual reports explaining what happened, the location where the incident took place, and the date and time when it happened. An NGO in the form of a foundation[20] registered both in the US and India has created and piloted the platform that since its launch has collected over 45,000 reports in India, Nepal, and Kenya and is expanding its usage worldwide, with 10,000 reports uploaded in 2020-21. The collected data are thought of to be useful to not only city dwellers but also the administration, NGOs, and police officials. This practice achieves multiple goals from creating awareness on harassment so that women and other disadvantaged communities are encouraged to break silence and to report harassment experiences to community building and leadership. Also, a data gap is filled in, by generating geo-localised datasets that local administrations (but also managers at large institutions such as university campuses) can use to address the problems. Users of the app can also make informed decisions on where to travel, at what time, and by which transportation. In terms of impact on local policies, Safecity succeeded to

[19] https://www.safecity.in/

[20] https://www.safecity.in/wp-content/uploads/2021/09/Red-Dot-Foundation.pdf

influence police patrol and local vigilance schedules, to augment budgets for CCTV surveillance and speed up decision-making to fix street lightning and toilets in areas perceived as unsafe (Bencke, 2019[21]; Adams, Lea & D'Silva, 2021[22]).

It is worth stressing that anonymous reporting is encouraged on the app and platform, and validity concerns are only very rarely raised in the validity checks that are routinely performed to test the reliability of the platform, via patterns' reliability.

A similar initiative is developed with a different business model and technology, branded as Safetipin[23]. Also located in India (9 cities) with more use-cases (21) in several African and Asian cities too, Safetipin is a social enterprise that has partnered with software designers and developers to devise a service having in mind city administrations as core stakeholders to buy and implement the model, along with citizens. The team offers on the first place a "safety auditing" service, to assess how an area, a neighbourhood, or a city are safe according to a rubric of indicators devised by the company team and an expert advisory board, encompassing lighting of streets, visibility, diversity, (number of people from different age group and gender present in a certain time), crowd, availability of public transport, walk paths, security, i.e., available police and other private security, openness, i.e., if the user has a proper view of the area properly or not, and feeling, i.e., how safe the app users report they feel at a given time in that place.

By "quantifying" safety levels of a given public area attributing a score to it, it becomes possible for policy-makers and implementers to measure improvements. The scope is broader than safety reflected as freedom from harassment as it also includes criteria to measure walkability, accessibility, and functionality of public spaces, which are all taken into account. The Safetipin Site app can be customised for each project to collect additional information required by governments. This could involve accessibility of bus stops, functionality of public toilets, last mile connectivity from metro stations, and so on.

The MySafetipin mobile app is thus made available to citizens who can use the multiple functions such as providing feedback about how safe they

[21] Bencke, F. (2019).Gender and The City. Building feminist geography approaches to public life surveys in urban India. University of Copenhagen, Master Thesis in collaboration with Safetipin/Red Dot Foundation.

[22] Adams, A., Lea, S.G. and E.M. D'Silva (2021). Digital Technologies and Interventions Against Gender-Based Violence in Rural Areas, International Criminal Justice Review. 2021;31(4): 438–455. doi:10.1177/10575677211040413

[23] https://safetipin.com/

feel in a space/area and "audit" the place themselves including pictures if they wish. All data provided by different means (both stakeholders and individuals allow to compute a safety score for a location, which can be then used by people to make safer choices by means of selecting a place to stay when travelling based on the safety score of the neighbourhood, driving or walking using the safest route, finding the nearest safe place and heading there. Advanced functionalities enable users to ask friends to track them and get notifications should they have any troubles or receiving notifications when in an unsafe place.

At the same time, stakeholders can be served through an extra function defined as 'Safetipin Nite' for collecting data both at night and during the day: by installing a phone with the app on the windshield of a car, routes can be worked out in back-end mode by trained professionals so that all streets of the city are covered: as the car moves, photographs are taken at predefined distances and are uploaded onto the platform servers. Machine learning (computer vision) algorithms are used on these pictures to extract information on safety parameters. A few additional data points are then added by trained coders who use this information to audit a given point, and this information, all geo and time-tagged, adds on to all other data that are collected by different means (individual citizens included) to measure the safety levels of urban spaces. The complex technology that has been created by integrating different apps is described in more detail at the developer's website[24]. The use of Women Safety Audits leveraging on (both open data and crowdsourcing) technologies has been supported by UN Women (see UN Women 2019, p. 20 onwards); Safetipin has been highlighted as a good practice by UN Habitat and UK Aid agency. Yet, some studies point at flaws and criticisms in its functioning and at areas for improvement in, among others, data visualisation components of the technology, as well as privacy and data protection (Manazir, Govinda & Rubina, 2019[25]).

Similar to the two above-mentioned experiences and the related technologies, a European version has recently kicked off, in the shape of the Safe and the City project[26], run by a women-led SME and piloted in London. It was awarded a Greater London Authority "Green Deal grant" to scale the technology in 2021 and by a further grant from the Foundation for Integrated

---

[24] https://metadesignsolutions.com/case-study-safetipin

[25] Sharique Hassan Manazir, Madhav Govind, and Rubinan (2019). My Safetipin Mobile Phone Application: Case Study of E-participation Platform for Women Safety in India, Journal of Scientometric Res. 2019; 8(1):47–53.

[26] www.safeandthecity.com

Transport (FIT) as funding enabling social enterprises to grow their business and impact. The tool is offering a free personal safety navigation app by which users can plan their routes and share and rate their journeys while keeping safe as they receive notifications walking to any areas the police have flagged as a historical street for crimes. In addition, if any emergency occurs, a two-click access allows to quickly contact appropriate emergency services. Users are asked to report when they feel unsafe, so that the shared information also feeds into the geo-tagged mapping system. The service privacy and data protection policy is GDPR compliant, and as the app is free for final beneficiaries/individual users, the incentive to report and provide personal data/information consists of using the services with no fee. Also this SME profiles itself as mission-oriented and runs awareness raising campaigns against sexual harassment against women and LGBTQ+ minorities.

The three case studies presented in this chapter show interesting bottom-up approaches where the initiative is taken by civil society organisations and/or social enterprises, to design cloud-based IT solutions, even using integrated cutting-edge technologies (as it is the case for Safetipin and Safe and the City in particular), that rely on both user-generated data and data that is collected by the organisations in partnership with local and police authorities. The final purpose is to tackle societal problems and facilitate women and minority right to safety in public space and particularly during nighttime, which have been the subject of extensive studies and literature (i.e., Ceccato, 2014[27]). The cases can also be considered as illustrative of increasingly spread gendered approaches to inclusive smart cities design (URBACT, 2022[28]; Sangiuliano, 2017[29]; Nesti, 2019[30]). Data and privacy protection are in all cases provided by users upon the incentive of a free-to-use service, and in the Europe-based service (Safe and the City), their use is regulated according to GDPR compliance.

The key features include an attention to cover and fill data-gaps that are usually featuring urban data collection when it comes to societal issues that might be undervalued in mainstream policies and perceptions from

[27] Ceccato, V. (2014). Safety on the move. Crime and perceived safety in transit environments, Security Journal, 27, 127–131.

[28] https://urbact.eu/smart-cities-innovation-and-gender-equality

[29] Sangiuliano, M. (2017). Smart Cities and Gender: main arguments and dimensions for a promising research and policy development area. Paper prepared and published for the OHCHR (Office of the High Commissioner United Nation Human Rights).

[30] Nesti, G. (2019). Mainstreaming gender equality in smart cities: Theoretical, methodological and empirical challenges. Information Polity, vol. 24, no. 3, pp. 289–304, 2019.

minorities; along with an attention to broaden the scope from mere statistics and quantitative data to qualitative ones including reports from users, this is also in line with those studies that argue in favour of more inclusive and empowering forms of data-driven civic engagement and activism (D'Ignazio & Klein, 2020[31]; Costanza-Chock, 2020[32]).

16.4 Envisioning the Future of Citizens' Intelligent Cities and the Role of Citizen Engagement

Climate change has become a serious challenge and consequently a topic of discussion among European policy-makers for more than two decades now. The European Union has been largely recognised to be a leader in promoting climate action within its borders and beyond them, attracting attention towards global environmental problems and supporting key agreements related to the environment such as the Paris Agreement. In 2019, the European Commission presented the European Green Deal[33] as a new growth strategy that aims to transform the EU into a fair and prosperous society, with a modern, resource-efficient, and competitive economy with net greenhouse gases emissions in 2050. A renewed awareness of the urgency to tackle climate change impact and focus on environmental policies has also changed the definition of "smart cities": the use of digital infrastructures and solutions has been framed more than it was the case already as means to achieve climate neutrality goals. In fact, achieving 100 cities that commit to cut emissions by 55% by 2030 is the objective identified by the Mission Board for climate-neutral and smart cities,[34] in view of making Europe climate-neutral by 2050. Digital technologies play a key role here; so these policies are integrated with Europe's digital policy and initiatives (such as the new Digital Europe Programme) thought of as a convoy to accelerate the transition: high-quality connectivity infrastructures and digital environments that empower end-users, complying with GDPR and European ethical standards, are expected to grow and develop further.

---

[31] D'Ignazio, C. and L. Klein (2020). Data Feminism. A new way of thinking about data science and data ethics that is informed by the ideas of intersectional feminism. MIT Press.

[32] Costanza-Cock, S. (2020). Design Justice. Community-Led Practices to Build the Worlds We Need, MIT Press.

[33] https://ec.europa.eu/info/strategy/priorities-2019-2024/european-green-deal_en

[34] https://ec.europa.eu/info/research-and-innovation/funding/funding-opportunities/funding-programmes-and-open-calls/horizon-europe/eu-missions-horizon-europe/climate-neutral-and-smart-cities_en

Studies have estimated that ICT could enable a 20% reduction of global CO2e emissions by 2030, holding emissions at 2015 levels, and that the ICT sector's emissions "footprint" is expected to decrease to 1.97% of global emissions by 2030, compared to 2.3% in 2020. Furthermore, the emissions avoided through the use of ICT are nearly ten times greater than the emissions generated by deploying it and with substantial benefits to agriculture as well including water and oil-use decrease (GeSI, 2015[35]). Even taking into account more critical opinions in terms of the environmental impact of the ICT industry than the above-mentioned ones, and the demonstrated need for sector-based efforts and regulations needed to make sure that the greenhouse emissions of the sector itself are in line with the Paris Agreements (Freitag *et al.*, 2021[36]), we can assume that it can play a positive role in achieving solutions in the face of the climate crisis. Private and public investments are expected in projects that use automation through AI and robotics to disassemble and separate waste, to enhance stability, effectiveness, and safety of low (zero) carbon electric grids, as well as in systems that model extreme weather phenomena, by climate impact modelling for mitigation and adaption, also thanks to collection, better access, enhanced analysis of environmental "big" data, and a new generation of digital tools for citizen' engagement.

The Interim Report from the Mission (EC, 2020[37]) identified the so-called Climate City Contracts as the key tools to be used by cities to set up programmatic objectives and strategies towards de-carbonisation for energy, transport, buildings, and even industry and agriculture. The mission policy report acknowledges that beyond R&I, a broader involvement is required across a spectrum of actors and policy areas and that a meaningful engagement of citizens is needed. Even if in the main part of the text and the dedicated chapter on citizen engagement of the report the formulation is not specific and in-depth as one could expect, the proposed framing for citizen engagement in smart cities that is presented in one of the reports' annexes is much more to the point and very close to the approach we have pursued within the Citizen Focus Action Cluster of the Smart Cities Marketplace, with a call to substantial impact, differentiation

---

[35] GeSI (Global eSustainability Initiative). #SMARTer2030. ICT Solutions for 21st Century Challenges.

[36] Freitag, C., Berners-Lee, M., Widdicks, K., Knowles, B., G.S., Blair and A. Friday (2021). The real climate and transformative impact of ICT: A critique of estimates, trends, and regulations. Patterns, Volume 2, Issue 9.

[37] European Commission, DG R&I (2020). 100 climate-neutral cities by 2030 - by and for the citizens Interim report of the mission board for climate-neutral and smart cities. Luxembourg, Luxembourg: Publications Office of the European Union.

from consultative/communication efforts, and inclusiveness: "To be effective, citizen engagement has to be *inclusive, deliberative, and influential.* These three basic criteria put it in a different league compared to other ways in which we reach out to society such as communication, public consultations or stakeholder debates. Citizens bring original perspectives to R&I and policymaking, and their engagement helps bridge the gap between science, markets and society. This is especially important in fundamental transformations – e.g. the transition to climate neutrality – that require not just innovation in technologies, but also profound changes in lifestyles and behaviour, along with innovative governance models. Such transformations cannot be imposed from the top: they need to be embraced and shaped by the citizens themselves" (ibid, EC, 2020, p. 23).

This policy framework and its objectives are supported by the Net-Zero Cities Consortium[38]that is set up to constitute a one-stop-shop that will offer support to cities committed to implement the climate contracts via open calls for proposals. Citizen engagement is a crucial part of the process for creating a Climate City Contract, along with the concept of co-creating the contracts with local stakeholders and citizens, since the design phase of the contracts and along the implementation. Following the open calls run in 2022 with the participation of 377 cities, the 100 cities that will be part of the programme have been selected.[39] Nevertheless, the EC has announced that the NetZero Consortium will also offer forms of support to the other applicant cities although via a twinning programme. The Net-Zero platform is therefore a digital platform to follow up in the next years when the challenge will be to have a more focused approach towards citizen engagement within climate neutrality policies in urban contexts.

Besides the fact that citizen engagement as a field will be much more oriented towards specific policy areas and topics related to climate change, we can also expect that the technological developments already mentioned above and reflected in case studies in paragraph 16.3.1 in particular and partly 16.3.2 will heavily influence this field of studies and practice. The increasing share of R&I endeavours in big-data analytics through artificial intelligence algorithms and systems in particular has vast potential applications for civic-engagement purposes as well (Brandusescu & Reia, 2022[40]): public trust

[38] https://netzerocities.eu/

[39] https://op.europa.eu/en/publication-detail/-/publication/822ee360-c9bf-11ec-b6f4-01aa75ed71a1/language-en/format-PDF/source-256649647

[40] Brandusescu, A., & Reia, J. (Eds.). (2022). Artificial intelligence in the city: Building civic engagement and public trust. Centre for Interdisciplinary Research on Montréal, McGill University. https://www.mcgill.ca/centre-montreal/projects/completed-projects/ai-city

and meaningful civic engagement can flourish and persist as data and artificial intelligence become increasingly pervasive in our lives, provided that some inherent challenges are critically approached and researched upon. In fact, these technologies pose new challenges both to core individual values such as privacy and equality, fairness, security, and accountability that have been outlined in studies (Gebru, 2020[41]) and tackled in the White Paper on Artificial Intelligence – A European Approach to Excellence and Trust[42] that require further investigation, research, and experimentation. Horizon Europe has integrated these concerns as research priorities under different clusters so that several calls for proposals have been published and probably more will follow on these issues: between the years 2022 and 2023, at least six new projects will be launched that will investigate, on one hand, how to de-bias AI systems and re-think fairness to avoid gender and intersectional bias is reproduced and multiplied in different domains[43], and, on the other, on how to harness the AI technologies for inclusive and fair civic engagement and democratic participation[44]: the generated research and the pilots that will be set up will offer knowledge and inspiration to advance this field for sure.

[41] Gebru, T. (2020). Race and Gender, in Dubber, M. D., Pasquale, F. and S. Das (eds.) (2020).The Oxford Handbook of Ethics of AI, pp. 253–270. New Your, Oxford University Press.

[42] Link to the White Paper on Artificial Intelligence.

[43] https://ec.europa.eu/info/funding-tenders/opportunities/portal/screen/opportunities/topic-details/horizon-cl4-2021-human-01-24 (check the list of funded projects to the bottom of the page)

[44] https://ec.europa.eu/info/funding-tenders/opportunities/portal/screen/opportunities/topic-details/horizon-cl2-2022-democracy-01-01 (the list of funded projects will be published to the bottom of the page when available after the Evaluation results will be published and the Grant Agreement signed with the winning consortia).

17

What Next?

Paolo Boscolo[1], Michael Mulquin[2] and Shaun Topham[3]

[1]Comune di Prato, Italy
[2]MIMs Ambassador, Open & Agile Smart Cities
[3]Lead, EU Smart Cities Marketplace, "Citizen's Control of Data" Initiative

Abstract

This chapter covers what action needs to be carried out to gather together solutions to the issues which were discussed, in order to make progress towards having a European model for the data economy. Follow-up activities are suggested and these would focus around establishing a core group of cities to collectively act as a lighthouse. There will be a focus on how to introduce personal data into the smart city data mix, in order to meet aspirations to become climate-neutral by 2030. Determining the value of data, through developing the "story of data within a smart city", would be one such activity. This, in turn, would help to answer some of the unresolved issues around producing business models for data platforms, whilst having a raft of exemplars and guidance coming from the projects contributing to this book to build upon and to be able to utilise the tools they have developed. The overall goal is to create the critical mass of citizens who would be willing to share their personal data within a smart city, which would give a boost to local data economies and make the likelihood of having a European model dominating. And, of course, the smart cities themselves would benefit with improved services and better informed decision-making and potential revenues for themselves and the local data economy.

17.1 Moving Towards a European Model for the Data Economy

Currently, the dominant non-European data economy model has a commercial focus and is not designed to maximise social and economic impacts for

DOI: 10.1201/9781003399384-18

a city and its citizens. In this model, cities may even have to purchase back their own data, whilst the personal data of its citizens also risk to be exported outside Europe, for corporate gain.

This is clearly recognised in all that is being done by the European Commission to hasten the emergence of a true European data economy model in line with EU values that take account of social and public needs.[1]

"As data becomes the new fuel of the economy and a key asset to address our societal challenges, the EU cannot afford to have the data of its businesses, public sector and citizens stored and exploited largely outside its borders. This is affecting not only our economic performance but also our security, safety and sovereignty. As announced in the EU data strategy, the EU has the means to become the world's most secure and trustful data hub".[2]

17.2 The Focus for Follow-up Activity

If we are aiming to radically change the existing situation, all the available tools must be utilised with regard to making a citizen's personal data more accessible and able to help realise the European vision.

Many of these tools have been described in previous chapters. Many more will be emerging from current and planned projects. Solutions and approaches are being suggested from legal, governance, interoperability, and business model perspectives.

Whilst having no pretensions of offering a joined-up and all-embracing solution to all the necessary requirements, we can suggest a starting point. In the tradition of developing lighthouse projects, a group of cities should gather to compare notes, share experiences, and move forward together as a team, to tackle all the issues focused upon in this book, along with other emerging issues.

- It should fit within and be helped by all the existing supporting activities and networks.
- It should act not simply as a guide to other cities but as a mechanism for them to join and share the value that such a group should be able to create, with the right conditions in place.

---

[1] "Communication: A European strategy for data," European Commission, p. 35, Feb. 2020, Accessed: Jul. 25, 2022. [Online]. Available: https://ec.europa.eu/info/sites/default/files/communication-european-strategy-data-19feb2020_en.pdf

[2] "European data strategy - European Commission." https://ec.europa.eu/info/strategy/priorities-2019-2024/europe-fit-digital-age/european-data-strategy_en (accessed Jul. 25, 2022).

- It should come up with answers as to how revenues may be generated and shared to drive forward the agenda.
- It should determine the most appropriate governance arrangements for such a grouping.
- It should push business models towards achieving the wider goals necessary for achieving climate neutrality.
- And it should strive to utilise the personal data of its citizens, to increase citizens input into the development of their smart and climate-neutral cities.

17.3 The Story of Data

We have covered the local data economy and finance platforms and elaborated on the evolution of business models within a smart cities and a data ecosystem context. But, all is built upon sand without a better understanding of what data, as a non-rival good, is actually worth. Chapter 14 goes into considerable depth regarding the problems faced in attempting to put a value on the use of data, in the midst of a wide range of existing and often contradictory ways of doing this.

As Safe-DEED concluded: "Finally, when reporting the value of data, we recall the observations made of Slotin *et al.* with respect to the efficacy of impact-based approaches to data valuation[3].The success of these approaches consists in the fact that they are able to tell compelling stories based on data and connect them to clear outcomes and contexts. This is also echoed by the Data Narratives approach,[4,5] which acknowledges that "the value of big data is not data, but the narrative that it generates and supports".

Safe-DEED recognised that this entire narratives approach is very interesting. "Starting from the story (the communication) itself, and define the information needs, which in turn defines the kind of analyses that can be

[3] J. Slotin, "What Do We Know About the Value of Data?" Accessed: Jul. 25, 2022. [Online]. Available: https://www.data4sdgs.org/sites/default/files/services_files/Value%20of%20Data%20Report_Final_compressed_0.pdf

[4] K. J. Hammond, "The Value of Big Data Isn't the Data," Harvard Business Review, 2013, Accessed: Jul. 25, 2022. [Online]. Available: https://hbr.org/2013/05/the-value-of-big-data-isnt-the

[5] M. Bergdahl et al., "Handbook on Data Quality Assessment Methods and Tools," EUROPEAN COMMISSION EUROSTAT, 2007, Accessed: Jul. 25, 2022. [Online]. Available: https://millenniumindicators.un.org/unsd/dnss/docs-nqaf/Eurostat-HANDBOOK%20ON%20DATA%20QUALITY%20ASSESSMENT%20METHODS%20AND%20TOOLS%20%20I.pdf

performed with the facts at hand. Finally, the required facts define how you are going to derive these elements of information from the data you have".[6]

Attention was drawn to the fact that "Data wrapped in stories are 22 times more memorable than bare facts".

Safe-DEED also recommended that we need to focus also on "reporting the results either by well-documented subsequent aggregations, by implementing a labelling/certification system"[6]. A common and joined up way of understanding the value of data within the smart cities environment is well worth pursuing. It was pointed out that research focused specifically on contexts for data value being almost non-existent.

Elaborating this "Story of data", working back from why the data is required, will also have the effect of educating the demand side of the equation. Having a description of the need for data in an understandable way from the point of view of service deliverers particularly can help to add the "Why?" regarding the gathering of data. It would promote "buy-in" at the city level and it would give a process to enable individual cities to make a start and focus on the elements of most interest to them, allowing for bespoke solutions as opposed to one size fitting all.

The situation we would like to get to would be a common cooperative ground with a more focused approach, with a stronger focus on net zero. This would co-exist with the current re-focusing of such initiatives as the Smart Cities Marketplace to adjust to the demands of the 100 cities approach. How can adding personal data into the mix support the planning for carbon-neutrality? A start would be in telling the compelling stories and connecting data to contexts and clear outcomes, highlighting the impact of good or bad data.

We have looked at the teams that would need to be assembled from a wide range of expertise in order to fully utilise personal data in the overall data mix. Mihnea Tufis adds to the list in Chapter 14, teasing out the more precise skills which would be required, whilst pointing out that by cities collaborating, resources need not be duplicated. He further added seven key recommendations for cities to follow, as well as a range of tools to start the process.

17.4 Business Models

Chapter 11 covered the evolution of business models in this area, with other chapters covering the local data economy and financial platforms.

[6] M. Tufiş, "D4.3 Report on context-aware and context-unaware valuation Status Final," 2020, Accessed: Jul. 25, 2022. [Online]. Available: www.safe-deed.eu

If we are to succeed in having a European model, with data and associated revenues and social value remaining in Europe, then there is a need to scale up. A critical mass of citizens will need to agree to share their data, under their own conditions, securely and with reward where that may be expected.

RUGGEDISED[7] made some interesting contributions to this chapter and included elucidating on the role of the network manager for a data platform. A key additional function described was the role of growing the ecosystem. "Ecosystem nurturing capability is the ability of the platform managers to nurture adoption and use of the platform as well as on-going collective innovation and exploration of new business models for the growth of the platform ecosystem".[8]

Further, the suggestion as to how to grow this network was made. "A lesson to be learned when dealing with attracting personal data to a platform is the requirement to reach a critical mass. Final revenues for a platform are equally dependent on a successful recruitment strategy which could entail cross-subsidising, with a pricing structure to support this".[9,10,11]

"It is expected that the establishment of the right pricing structure by the platform manager influences the adoption decisions of platform users and supports network effects within a platform ecosystem".[9,12]

Just the 100 climate-neutral cities represent 75 million people, well over 10% of the EU population.

Success in moving towards a critical mass will require the kind of governance arrangements discussed in Chapter 14, covering the governance

---

[7] "RUGGEDISED - Smart city lighthouse project." https://ruggedised.eu/smart-solutions/smart-solutions-overview/ (accessed Jul. 22, 2022).

[8] M. A. Jeusfeld, C. Quix, and M. Jarke, "Design and analysis of quality information for data warehouses," Lecture Notes in Computer Science (including subseries Lecture Notes in Artificial Intelligence and Lecture Notes in Bioinformatics), vol. 1507, pp. 349–362, 1998, doi: 10.1007/978-3-540-49524-6_28/COVER

[9] P. B. Crosby, "Quality Is Free: The Art of Making Quality Certain," 1979.

[10] F. de Amicis, D. Barone, and C. Batini, "An Analytical Framework to Analyze Dependencies Among Data Quality Dimensions.," in Proceedings of the 2006 International Conference on Information Quality, ICIQ 2006, Jul. 2006, pp. 369–383.

[11] T. Gebru et al., "Datasheets for Datasets," Commun ACM, vol. 64, no. 12, pp. 86–92, Mar. 2018, doi: 10.48550/arxiv.1803.09010.

[12] J. P. Carrascal, C. Riederer, V. Erramilli, M. Cherubini, and R. de Oliveira, "Your Browsing Behavior for a Big Mac: Economics of Personal Information Online," in Proceedings of the 22nd International Conference on World Wide Web, 2013, pp. 189–200. doi: 10.1145/2488388.2488406.

of personal data for the public interest being discussed and implemented according to local conditions but also fitting into a wider structure to reap the benefits of scale. In an ideal situation, this would include a discussion on how to utilise any revenues generated in addition to the benefits accruing in fulfilling the requirements established by a city.

With such a growing ecosystem, the local data economy would have a stimulus and the opportunity to flourish. But this would require the technical support for the wide range of tools used in developing such an ecosystem. "From a technical perspective, a platform should also provide business model support tools to enhance the economies of scope by encouraging new communities (e.g., data-driven start-ups, developers, and established firms) to join the platform ecosystem in order to explore new business opportunities, or to enhance their existing business models".[8]

Having the technical support tools such as application programming interfaces and software developer kits allows access and interaction with the platform and mediates between the platform and its users and so helps the local data economy by opening up new business opportunities within the ecosystem. Most of the projects referred to will have their own repositories of manuals and documents to support their software to smooth take-up.

17.5 A "Personal Data-Smart Cities" Group

Such a group would seek to provide a focus for primarily demonstrating all the best practices in utilising personal data in a city, aided by the existing support mechanisms, which would result in a focused joined-up approach by default, given the core focus for the group of cities which would comprise the group, but also providing a mechanism to take the lessons forward, enhanced with the new inputs from the support initiatives. This would have an aligning effect and potentially be self-financing.

The European Commission, in establishing the larger group of "100 Climate-neutral Cities by 2030", aims to exploit the cities' access to capital, know-hows, and economies of scale for the development to pilot and scale up green innovations. Green initiatives can be piloted and these cities will function as an innovation hub and showcase for the rest of Europe to follow and become climate neutral by 2050.

The "Personal Data-Smart Cities Group" would be able to contribute with a "specialist knowledge" to help achieve these aims.

An example of an approach that could be adopted is that of Viable Cities, which is developing new form of initiative to drive systems innovation for transformation in line with the mission of achieving climate-neutral cities by 2030.[13]

"A system demonstration is a controlled method for testing sustainable systems transition consisting of a combination of innovative solutions in a real-world environment –and with consideration for the context in which it is intended to function. This involves working with multiple levers of change (business models, regulation, forms of governance, behaviour, technology, etc.) based on a specific geographical environment or defined challenge area, in order to learn and build a portfolio of interventions that help to learn and change real-world systems. This is always with the aim of accelerating the climate transition in cities and co-benefits from climate transition such as improved health, more jobs and new business opportunities. It is, not least, a matter of mobilising investment and creating opportunities to scale up and disseminate the measures that prove effective in making the climate transition happen".[13]

For example, an experimentation based around the "story of data" would fit such an approach.

Such an initiative would not seek to replicate or replace anything which currently is serving the purpose of supporting the drive towards making EU cities carbon neutral. But a focal point whereby a small number of cities will specifically look at the issue of how best to utilise a citizens personal data in order to drive forward and enhance urban platforms and digital twins, etc., will generate a specific resource for the supporting networks to draw on.

Net-Zero Cities[14] is a project supporting the European Union's Green Deal[15] and has been designed to help cities overcome the current structural, institutional, and cultural barriers they face in order to achieve climate neutrality by 2030. In this capacity, it is supporting the "100 Climate-neutral Cities".[16]

It will help European cities by providing them with the support and solutions they need to achieve their Net-Zero goal in a socially inclusive way. The "Personal Data-Smart Cities" would be able to complement and contribute and benefit from this work.

---

[13] "Viable Cities." https://en.viablecities.se/# (accessed Jul. 25, 2022).

[14] "Net Zero Cities." https://netzerocities.eu/ (accessed Jul. 25, 2022).

[15] "A European Green Deal - European Commission." https://ec.europa.eu/info/strategy/priorities-2019-2024/european-green-deal_en (accessed Jul. 25, 2022).

[16] European Commission, Directorate-General for Research and Innovation, Gronkiewicz-Waltz, H., Larsson, A., Boni, A., et al., "100 climate-neutral cities by 2030 - by and for the citizens : report of the mission board for climate-neutral and smart cities", Publications Office, 2020, https://data.europa.eu/doi/10.2777/46063

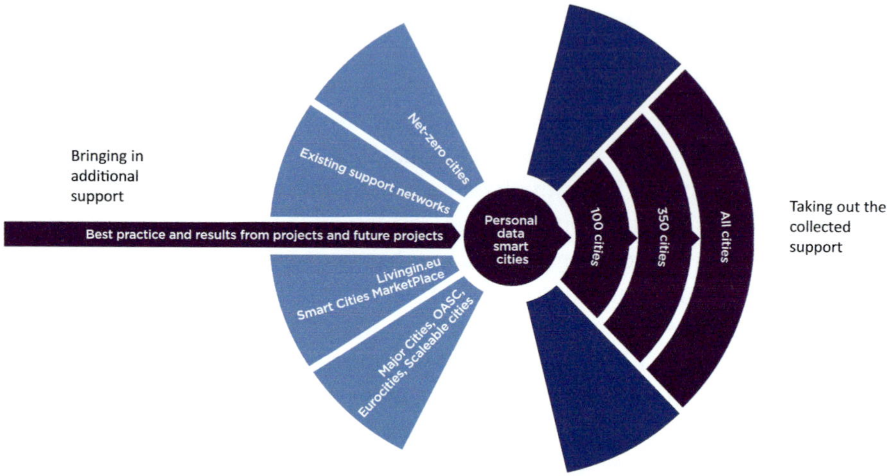

Figure 17.1 Role for a group of personal data-smart cities.

All the various supporting activities would become more valuable if they were to be focused on creating an excellent demonstration platform for the use of personal data and, in turn, would reap the benefits of the outcomes of this activity to use in their work into the future.

17.6 Citizen Engagement

Operational approach: Citizen's engagement should be an objective from the start as citizen use and acceptance supports the overall value creation for urban data platforms. Research in RUGGEDISED has not clearly identified when to include citizens, but it is encouraged to plan that engagement – whenever it may appear – from the beginning. Chapter 16, "Data-driven and Citizens' Inclusive Smart Cities", has focused on the citizen's role, whilst there are examples in Chapter 4 on how current uses of personal data can improve citizen's input into running a city. This is a topic to be further refined and improved in order to make a valuable tool. Simple consultation processes can be enriched by encasing them with additional personal data.

Regarding greater citizen participation, the previous chapter drew attention to the Net-Zero platform being a digital platform to monitor in the coming years to help meet the challenge of having a more focused approach towards citizen engagement within climate neutrality policies in urban contexts. And, further, several new projects will be announced to tackle issues arising from the development of new technologies, challenging individual values including privacy and accountability, equality and fairness, whilst looking at how to harness the technologies for inclusive and fair civic engagement and democratic participation, with the demonstration activities worthwhile following.

Chapter 10 looked at what is in the pipeline and what should be in the pipeline regarding the standards that need to be in place for citizens to have a consistent experience and not to be side-lined. A view on further needed actions was provided to enhance citizen services, including the data aspects, citizen-oriented management of local authorities, and citizens' security.

17.7 Governance

The first chapter in this volume sets out methods to steer future technical developments in the direction of Utopia and away from Dystopia. The first of four conditions identified is in the process of being met and of having the right privacy-enhancing technology that respects human rights and ensures responsibility. A steer had been given in Chapter 13 to the governance models suited for taking the direction towards "Utopia" in having "The right governance to implement the strategies that mitigate the 'regression to dystopia'" along with a set of recommendations towards having the right leadership, ethics and values from citizens and institutions to obliterate the digital divide and safeguard human rights. Having the right research to fully understand digitalisation and the behavioural, economic and political dynamics of our augmented humanity is needed whilst having a group of cities willing to be at the forefront of showcasing the outputs of this research can provide a focal point.

In Chapter 13, Marina Micheli indicated the models for governing the data ecosystems emerging, with one of them suggesting that local public administrations could play a key role in addressing power unbalances of the current data landscape acting as trusted data intermediaries and enabling the use of citizens' personal data for the public interest. She pointed out that such a model was still at a prototype stage, but the potential for utilising it is there, acting as a source for inspiration on how best to progress in the public interest. A recommendation was to "form or join alliances between cities which would enhance their opportunities to access, use and better govern (personal) data for the public interest".

Attention was paid to outlining the new skills and roles which would be required to overcome current barriers to progress, one of which is the emergence of "data stewards" in cities. This was initially a recommendation from the high-level expert group on B2G data sharing and described "individuals or teams that are empowered to proactively initiate, facilitate and coordinate" data sharing.[17] Their role would be to systematise data partnerships and scale efforts; hence, they will have the expertise for promoting data access, sharing, and management.

This would be consolidated by incorporating the job-description of such a facilitator as set out in Chapter 11, in which the need for managers to also look at growing an ecosystem and realising the potential of reaching a critical mass of cities collaborating and subsequently producing revenues to escalate the process.

With regard to the suggested group of cities, ideally, a team would consist of these managers as well as representatives of those working within a public administration and assisting the leadership in the disciplines which this book has attempted to make a start in covering. They would be lawyers, economists, service providers, strategists, technicians, and, of course, those in control of the emerging data ecosystems.

Peer-to-peer relations of officers at a similar level have been shown to be productive. Teams from a variety of disciplines, collectively addressing the issues, may help remove obstacles before they present themselves. But these teams would also benefit with academic participation to capture the lessons to be learned and subsequently to be shared by participating in the process as it evolves as active members, and subsequently contribute to research. The intention would not to be having a "possessor" of a good practice passing it down to others but for all to move forward collectively.

17.8 Interoperability

Interoperability is crucial if an ecosystem is to flourish and a critical mass of cities and citizens are to be involved. Whilst Chapter 7 covered the MIMS approach, Chapter 8 looked at interoperability within the health sector in a city, and Chapter 9 described a case study for demonstrating the MIM4 approach to interoperability between several existing projects and initiatives, including DataVaults, KRAKEN, and MyData. The context is set for local data sharing ecosystems, where data from many different agencies can be brought together to enable the city to be managed in a more holistic way. It points out that this requires technical, information, and organisational interoperability and provides a list of some of the specific areas where interoperability is needed in such an ecosystem. It then places this

---

[17] European Commission, 2020. Towards a European strategy on business-to-government data sharing for the public interest. Final report prepared by the High-Level Expert Group on Business-to-Government Data Sharing, at https://op.europa.eu/en/publication-detail/-/publication/d96edc29-70fd-11eb-9ac9-01aa75ed71a1

within the European Policy Context. The concept of minimal interoperability is then dealt with as a way of enabling small- and medium-sized cities and communities to put in place "good enough" interoperability mechanisms to enable effective data sharing without requiring excessive time or resources to implement. The chapter closes by reviewing the minimal interoperability mechanisms being developed by Open and Agile Smart Cities that are incorporated within the Living-in.EU initiative, thus helping to bring consistency into the marketplace.

However, there should be no limit as to how many other projects and initiatives could join in such experimentation, supporting the development of a critical mass of users, with its benefits to any business case.

When considering why you would want to have a project covering the use of the personal data from a football club's supporters to be compatible with a project dealing with the logistics within a port, there may not be an obvious answer. But if you add into the equation, tackling the problems of traffic in a city, then having the experiences from the travellers going to a match, alongside all the data about lorry movement, etc., then a local traffic strategy could benefit. But, looking at it from a perspective of a growing ecosystem, most cities on the list of 100 carbon-neutral cities will have a football team and many of them are ports. Consolidating the use-cases will make all more attractive and increase the likelihood of gaining a critical mass of users. A talented designer would have work to do in illustrating such a snowball effect!

17.9 Legality

It is also clear that the potential legal obstacles are being tackled. As Chapter 15, covering the legal and ethical aspects, pointed out, vast reforms are underway and an update of the European regulatory landscape was announced in terms of the Commission's Mission Statement for 2019–2025. Especially, some of them are expected to be significant for the deployment and use of personal data platform. The EC aims at renewing its overarching framework to achieve the proper balance between the wide availability and use of data with the high preservation of privacy, security, safety, and ethical standards. Aspects related to data ownership and data governance are going to be addressed and/or reframed. The strategy is motivated by the need to put people first in developing technology and to defend and promote European values and rights in how the technology is designed and deployed in the real economy and it sets out a programme of policy reforms, already started with the Data Governance Act, the Digital Services Act, the Digital Markets Act, and the Cybersecurity Strategy.

17.10 On the Horizon

2030 is no time away for the 100 cities to become carbon-neutral. Meeting this ambitious target will also mean the adoption of the emerging technologies, be they related to energy, mobility or in our case optimising the usage of data of all kinds, augmented by personal data, will also mean a requirement to rapidly introduce these technologies, immediately they are functional. Any delays which have often occurred in the past between the development of technology within a project and its eventual deployment need to be reduced in the current circumstances.

Whilst this work has concentrated on bringing the results from a wide range of projects having just finished or are due to finish or having results available to share, we must take cognisance of what is on the Horizon. Given the urgency of requiring drastic action quickly in order to achieve net-zero, we cannot afford to build into the time schedule any delay between a project delivering their findings and then these results taking a further period of time before they are deployed.

For example, it is necessary to monitor and engage with the projects announced as a result of the call for projects under the Horizon Programme Call "World Leading Data and Computing Technologies 2022 (Horizon-CL4-2022-Data-01)".[18]

The emphasis is on the development and demonstration of practical and mature end-to-end systems, building on the results of work on data platforms (topic H2020-ICT-13-2018-2019), privacy-preserving technologies, and computing technologies under Horizon 2020 and this programme. Much of this work on data platforms was carried out by projects referred to in this work.

The new projects about to start late in 2022/early 2023 are expected to contribute to the following expected outcomes:

> "improve the digital technologies, solutions and interoperable frameworks for data markets and data economy (e.g. industrial, administrative and societal/cultural data platforms/data spaces), allowing for data assets to be discoverable, efficiently and fairly priced and shared/traded in a secured, user-friendly, compliant and energy-efficient

[18] "Funding & tenders. Technologies and solutions for data trading, monetizing, exchange and interoperability (AI, Data and Robotics Partnership) (IA) - TOPIC ID: HORIZON-CL4-2022-DATA-01-04."https://ec.europa.eu/info/funding-tenders/opportunities/portal/screen/opportunities/topic-details/horizon-cl4-2022-data-01-04 (accessed Jul. 25, 2022).

> way; promote the development of a European industrial ecosystem of the data economy capable of ensuring digital autonomy; develop training material to endow workers in this occupational group with the right skillset in order to deploy the new technologies".[17]

The focus of these new projects is on technologies, solutions, and frameworks that facilitate the collection, sharing, storing, processing, trading, and re-using of data in compliance with the legal framework and satisfying the needs, expectations, and rights of the data providers, brokers, users, and data subjects.

They have to create practical and scalable solutions for handling large amounts of transactions while minimising energy consumption is necessary (e.g., smart and automated contracting, data rights management, and tracking of subsequent data use). They will be paying special attention to fostering approaches that ensure data and metadata interoperability, including the application of appropriate standards, reference architectures, common ontologies, vocabularies, and data models allowing smooth data sharing and this also across sectors.

These projects will also be expected to develop and support data spaces of realistic scope and size, deployable in real-world applications in various application areas. In particular, the actions are expected to support the deployment of the Common European Data Spaces, covered in "section 2.2 Data for EU" under the Digital Europe Programme for 2021–2022.[19]

A mechanism will be the Data Spaces Support Centre, which is to be set up under the Digital Europe Programme. All the projects will have an obligation to engage with a wide range of European data sharing schemes.

All this activity is expected to lead to the EU having world leading data and computing technologies. And closely linked to the goal of 100 carbon-free cities is the ambition to make Europe the first digitally led circular, climate-neutral and sustainable economy through the transformation of its mobility, energy, construction, and production systems, which will also depend heavily on having a globally attractive, secure, and dynamic data-agile economy. There is an argument for following the progress of this ongoing work, rather than to wait for it to be completed.

---

[19] European Commission, "ANNEX to the Commission Implementing Decision on the financing of the Digital Europe Programme and the adoption of the multiannual work programme for 2021 - 2022," 2021, Accessed: Jul. 25, 2022. [Online]. Available: https://eur-lex.europa.eu/legal-content/EN/TXT/?uri=consil%3APE_13_2021_INIT

17.11 Contracts to Have Data Plan

With the mission "100 Climate-neutral Cities by 2030", the European Commission aims to exploit the cities' access to capital, know-hows, and economies of scale for the development to pilot and scale up green innovations. 100 European cities have been selected and supported in their transformation towards complete climate neutrality by 2030. Here, green initiatives can be piloted and these cities will function as an innovation hub and showcase for the rest of Europe to follow and become climate-neutral by 2050.

These 100 climate-neutral cities will be developing contracts, based on guidance from the Commission. Swedish cities have had experience of this process, with Gothenburg,[20] as part of the Viable Cities movement, having produced an example of a "Climate City Contract". (Similar work has been ongoing in other countries such as Italy and Spain.) The initial overall strategic discussions underpinning the Swedish process were followed by a number of ambitious strategic initiatives, or system demonstrators, which give concrete form to the city's overall strategy of transition and create the conditions for success. These included progressing the work towards practically useful management and control tools for climate transition.

Their digital twins and a digitalised planning process are at the heart of this and a deep knowledge of management and governance challenges linked to climate transition has grown. But again, underpinning all this is the use of data, and particularly when considering issues of citizens participating in the process, consideration should be given in this contractual process as to how best to utilise a citizen's personal data. As Chapter 14 drew attention to, "For those cities that are still at the beginning of their digital transformation journey, this may be a good opportunity to do things right from the onset, by creating the data collection and management infrastructure, processing pipelines and metadata in a coherent way".

17.12 Concluding Remark

The foreword to this book expressed "The hope is that it will inspire smart cities to engage more actively in using relevant data, and in particular citizens' personal data, to support important local policy objectives, notably to become climate neutral as quickly as possible". And it is fitting to end with this.

---

[20] Viable Cities, "Climate City Contract 2030," 2021. Accessed: Jul. 25, 2022. [Online]. Available: https://static1.squarespace.com/static/5dd54ca29c917941 1df12b85/t/61bcccdf-336b2422ef7f0a98/1639763168501/Climate_City_Contract_2030_ENG_Goteborg.pdf

Index

D

About the Contributors

Ana García Garre, MIWenergia, Spain, graduated in industrial technologies in 2015 from the Polytechnical University of Cartagena. She has a master's degree in industrial engineering and is currently working toward her Ph.D. degree in demand response and energy efficiency. She is working as an Industrial Engineer with MIWenergia, participating as Project Manager of several European projects. ana.garcia@miwenergia.com

Borja Molina-Ríos, MIWenergia, Spain. He graduated in mathematics and computer engineering from the University of Murcia in 2020. Since then, he has been working at MIWenergia as part of the IT Department with involvement in different European projects. Recently, he finished the master's degree in big data analysis. borja.molina@miwenergia.com

Christina Tsiligkiri has been working for Olympiacos SFP since 2012, and since 2016, she has been holding the position of General Secretary being responsible for the communication, marketing, membership, and international relations' departments of the club. She has a five-year experience on running, coordinating, and assessing European projects co-funded by the European Union under the framework of Erasmus+. She has also studied accounting and finance in the Thessaloniki TEI, continued her studies with a one-year scholarship in Rennes (France) and, finally, graduated in 2009. ctsiligkiri.erasmus@olympiacos.org

Elena Palmisano received a degree in civil engineering in 1988 and the Ph.D. degree in methods and technologies for environmental monitoring in 1994 with the University of Florence (Italy). She works as a consultant with PIN Scrl (a public/private consortium managing the University of Florence site in Prato) and with the Municipality of Prato. She has taken part in many regional, national, and EU projects in the field of innovation technology, ICT applications in different contexts, such as info-mobility, e-government, cultural heritage, and education. elena.palmisano@gmail.com

Giannis Ledakis graduated from the Department of Computer Engineering and Informatics of the University of Patras. His Diploma Thesis was in the area of software design, creating a language for JavaScript and AJAX enhanced HTML forms creation, defined with XML schema usage. Since 2007, he has been working with UBITECH LTD as a Software Engineer in both commercial and research ICT projects, including many H2020 projects. He is a part of the management and development team of DataVaults. gledakis@ubitech.eu

Haydee S. Sheombar is an experienced international business consultant, academic researcher, and speaker working in the public and private sectors. As the founder of Kankan*Tree, she advises leaders, managers, and policy-makers on digital transformation. She has a research affiliation with Erasmus University Rotterdam. Her past experience includes over 20 years with PWC and IBM. She was IBM's first Benelux Smarter Cities Director.

Ilesh Datani, Assentian, Ireland. He has over 10 years of operational experience in data security and data privacy developing and delivering novel innovative solutions to support data-driven automated decision-making in financial services and more recently in the logistics sector. He is on the British Standards Committee for Big Data and the International Standards Committees for Information Security and Artificial Intelligence. He is a published author having contributed to books on digital finance and reports on cyber security and blockchain in healthcare.

John Ketchell counts over 30 years of activity in the standardisation world, principally in Information and Communications Technologies. Currently, he is the Chair of ANEC's Digital Society Working Group (ANEC represents consumer interests in European standardisation) and of a Working Group of the ETSI Board concerning how to improve the inclusiveness of standards activities. He represents ANEC on CEN's Technical Committee on smart and sustainable cities and communities.

Joseph V. Sheombar is a master's student with the Technical University of Delft's Architecture Urbanism and Building Sciences faculty. As the integration of technology in the built environment is the inevitable future, he has selected smart cities combined with real estate as the topic of his thesis. He completed a summer internship as a research assistant with Kankan*Tree.

Juergen Ritzer: Following a 15-year corporate career in national, European, and global business development and strategy positions, he founded a

European network consultancy on low carbon business models (2009) and became co-founder of classroom Global Energy MBA Investment Fund (2010) and co-founder of platform EEIP, operating a global information and engagement network of 150.000 users from business, policy, science, and civil society (2011). He acts as EEIP Business Director with specific focus on disruptive technologies and business models and the drivers for implementation at scale.

He co-chairs the Smart City Marketplace Action Cluster on Business Models & Finance. https://ee-ip.org

Konstantinos Oikonomou is a Software Engineer and researcher with Ubitech LTD. He received the Diploma in electrical and computer engineering from the University of Patras and the M.Sc. degree in advanced software engineering with management from the King's College of London. He has been working since 2016 in various projects including many H2020 projects and is a part of the DataVaults development team. koikonomou@ubitech.eu

Marina Da Bormida in Cugurra is a Lawyer and Ethics Expert, specialising in R&I projects, in particular in legal issues pertaining to information society and new technologies with a Ph.D. degree from the "Telematics and Information Society" Ph.D. School in University of Florence. She is also an expert in ethical and societal themes related to ICT research and technological developments. She is serving as independent Ethical Expert for European Commission and European Defense Agency. She has scientific collaboration with CNIT (National Inter-University Consortium for Telecommunications), CNR - ITTIG (Italian National Research Council, Institute of Legal Information Theory and Techniques), and INSIGHT Centre for Data Analytics - NUIG (National University of Ireland Galway). She is a Legal Advisor in the R&I Division of multinational companies. She contributed to the activities of the legal working groups of EU-wide initiatives (EU Blockchain Observatory Forum, BDVA/DAIRO Task Force "AI and Big Data for the Financial Sector") and Chair of the Ethics, Data Protection and Privacy (EDPP) Task Force of the "Citizen's Control of Personal Data" initiative within Smart City Marketplace.

María José López is a Computer Engineer with the master's degree in software engineering from the University of Basque Country and the master's degree in computational engineering and mathematics from the Rovira i Virgili University. Maria is a Senior Researcher with Tecnalia Research and Innovation Foundation (member of Basque research and technology alliance), focused on the area of data platforms and data spaces and is also a part of the Digital Transformation Unit developing new massive, heterogeneous data processing technologies and architectures on high-performance platforms. MariaJose.Lopez@tecnalia.com

Marina Micheli is a Scientific Project Officer with the Digital Economy Unit of the European Commission's Joint Research Centre (JRC). Her work explores data governance for the public interest, examining, in particular, the perspectives of social actors and the emerging governance models. She has a Ph.D. degree in information society from the Department of Sociology and Social Research of University Milano-Bicocca, Italy and her background is at the intersections of digital media studies and sociology.

Maria Sangiuliano, Research Director, Smart Venice, has an MA degree in philosophy and social sciences and a Ph.D. degree in cognitive and learning sciences. Maria has 20+ years of experience in coordinating (mostly EU funded) research and programmes studying socio-technical innovations and inequalities that leverage on participation, engagement, and empowerment, both at the organisational level and within territorial ecosystems. Her focus is on designing and implementing participatory action-research projects and capacity-building programmes that aim at promoting social justice, environmental sustainability, and equality.

She has been chairing the Citizen Focus Action Cluster at the Smart Cities Marketplace since 2015, an initiative jointly run by several DGs from the European Commission, and she is an expert member and contributor of the URBACT Gender Equal Cities network.

LinkedIN profile, email: maria.sangiuliano@smartvenice.org

Michail Bourmpos is the Head of IT Department, City of Piraeus. He received the Diploma in electrical and computer engineering from the Aristotle University of Thessaloniki, Greece in 2000. In 2001, he was granted the M.Sc. degree in control systems from Imperial College, London, UK. In 2002, he was granted the M.Sc. degree in advanced computing from Imperial College, London, UK. In 2005, he received the MBA degree in techno-economical systems from the National Technical University of Athens, Greece. He also received the Ph.D. degree in information technology and telecommunications from the National and Kapodistrian University of Athens in 2017. mmpour@piraeus.gov.gr

Mihnea Tufiş is a data scientist at the Data Science and Big Data Analytics group at Eurecat. Mihnea holds a joint M.Sc. degree in data mining and knowledge management from Polytech Nantes (France) and Eastern Piedmont (Italy) Universities, and a Ph.D. degree in computer science from the Sorbonne University - Pierre and Marie Curie in Paris. His Ph.D. research focused on using crowdsourcing to extract information from multimodal text and image media, the post-processing and analysis of collected data, and its storage in dedicated formats. He developed Comics++, a unique crowdsourcing platform for annotating comics. Since 2018, he has been involved in research and development projects in the areas of smart urban mobility, developing predictive models for traffic parameters estimation, and product delivery management.

Since joining Eurecat, he has been involved in a variety of research and innovation projects. His main current research is focused on how data value is connected to data quality and data exploitability. He is also leading Eurecat's new research line on responsible AI, bringing together R&D from the areas of privacy, data and algorithmic fairness, and explainable AI. Alternatively, he is participating in a range of technology transfer projects, implementing data-driven solutions for beneficiaries from various industries: cultural, urban mobility, human resources, and retail.

In writing this chapter, Minhea has received the support of the Safe-DEED project, funded from the European Union's Horizon 2020 research and innovation program under grant agreement No. 825225.

Sotiris Koussouris holds a Ph.D. degree in information systems and business processes and is an Electrical and Computer Engineer working on developing novel solutions around big-linked-open data, analytics and AI, and cloud computing for various verticals belonging to the industry as well as the public sector. Suite5 Data Intelligence Solutions Ltd. sotiris@suite5.eu

Valentin Sanchez Pelaez has a B.Sc. degree in physics with more than 20 years' experience in the analysis, research, and development in the ICTs field and more than 15 years' experience in project management. He is a Project Manager with Tecnalia Research and Innovation Foundation (member of Basque research and technology alliance), focusing on the area of data platforms and data spaces. He is a part of the Digital Transformation Unit developing new massive, heterogeneous data processing technologies and architectures on high-performance platforms. valentin.sanchez@tecnalia.com

Vincent Keunen is the founder and CEO of Andaman7. Previously, he developed numerous medical IT systems (a secure data exchange platform used by more than 90% of Belgian hospitals and family doctors, and a prevention electronic health record, managing today close to 1 million patients). Vincent was also the Chief Information Officer (CIO) of success story Lampiris and was awarded the CIO of the year prize for his work there. When Vincent was diagnosed with leukaemia in 2007, and his 10-year-old son was diagnosed with bone cancer 3 months later, he decided to use his software engineering skills to provide all patients with a tool to manage their health information on their smartphone. That platform can now be used to improve medical research by collecting RWE (real world evidence) and PRO (patient reported outcomes) while facilitating DCT (decentralised clinical trials), telemedicine, remote patient monitoring, and continuity of care.

Yury Glikman is the project manager and head of the Open Service Engineering group at Digital Public Services Competence Center at Fraunhofer FOKUS. In the past 18 years, Yury was actively contributing and was leading technical activities in numerous international and national research projects and in research projects in collaboration with Japanese industry. In the past, Yury was successfully coordinating the FP7 projects: OCEAN, Policy Compass. He is currently the coordinator of the H2020 IA PPP project AEGIS (Advanced Big Data Value Chains for Public Safety and Personal Security). yury.glikman@fokus.fraunhofer.de

About the Editors

Shaun Topham has a long experience working with government and public administration around the world, having also been a politician. He was the President of EU eForum, a network comprising cities, universities, and companies, initially established by the Commission to share best practice in eGovernment whilst focusing in recent years on linking Chinese and European smart cities and on identity management. He has been involved in piloting and disseminating numerous projects. Amongst them are STORK and STORK2.0 covering cross-border identity, and EKSISTENZ on identity theft. He was also a Fellow of the University of Manchester, Urban Institute. shaun.topham@eu-forum.org

Paolo Boscolo, He received the master's degree in electronics engineering in 1986. He has worked for Infogroup S.p.A. and Nuovo Pignone S.p.A, and from 1988 to 1991, he worked at ALCATEL Italia in Florence, participating in EU projects such as Race/TELEMED and Race/EPIN. In 1990, he became Head of the IT department in Florence for ALCATEL Italia. In 1992, he joined the Comune di Prato as Head of the Information Service, managing all distributed computing facilities and multi-protocol data transmission networks in the city. He participates in national innovation projects and in European Projects in the field of advanced IT infrastructures, digital innovation, and data exploitation. He is currently the IT Director of Prato City Council and responsible for the Innovation & Digital Agenda office and Coordinator of the Open Data publishing team. Since 2008, he has been representing the Prato City Council in the Major Cities of Europe (MCE) and is a member of the Executive Committee. p.boscolo@comune.prato.it

Michael Mulquin has spent nearly 30 years partnering with cities, rural areas, and industry on how technology can help neighbourhoods and cities work better. For the last 10 years, he has been focusing on the development and implementation of smart city standards. He is the Chair of the IEC Smart Cities Systems Committee and Co-Chair of the IEC-ISO-ITU Joint Smart

Cities Task Force. He is City Standards Associate with the UK Connected Places Catapult and a member of the UK Government External Advisory Group on Secure Connected Communities. He works with Open & Agile Smart Cities (OASC) to support the development of the minimal interoperability mechanisms (MIMs) that provide focused sets of requirements to support different aspects of building a local data sharing ecosystem. michael@oascities.org

For Product Safety Concerns and Information please contact our
EU representative GPSR@taylorandfrancis.com Taylor & Francis
Verlag GmbH, Kaufingerstraße 24, 80331 München, Germany